AF360823

AI in Education

AI in Education

Editor

Juan Cruz-Benito

MDPI • Basel • Beijing • Wuhan • Barcelona • Belgrade • Manchester • Tokyo • Cluj • Tianjin

Editor
Juan Cruz-Benito
IBM Quantum. IBM T.J.
Watson Research Center,
Yorktown Heights, NY 10598, USA

Editorial Office
MDPI
St. Alban-Anlage 66
4052 Basel, Switzerland

This is a reprint of articles from the Special Issue published online in the open access journal *Applied Sciences* (ISSN 2076-3417) (available at: https://www.mdpi.com/journal/applsci/special_issues/AI_in_Education).

For citation purposes, cite each article independently as indicated on the article page online and as indicated below:

LastName, A.A.; LastName, B.B.; LastName, C.C. Article Title. *Journal Name* **Year**, *Volume Number*, Page Range.

ISBN 978-3-0365-4341-3 (Hbk)
ISBN 978-3-0365-4342-0 (PDF)

Contents

About the Editor

Juan Cruz-Benito

Juan Cruz-Benito received his Ph.D. in Computing Engineering from the University of Salamanca, Spain, in 2018. He had previously completed his M.Sc. degree in Intelligent Systems, B.Sc. in Computer Systems, and Technical Engineering Degree in Computer Systems from the same university in 2013, 2012, and 2011, respectively. He is currently Research Developer at IBM Quantum, IBM T.J. Watson Research Center, NY, USA. Prior to this, he worked for five years in R&D projects at the GRIAL Research Group, University of Salamanca, Spain, where he participated as a software engineer, researcher, and developer in numerous European and Spanish R&D projects (public and private ones). His research interests involve human–computer interaction, machine learning, data science, and technologies for educational purposes. He is the co-author of more than 80 publications, serves as a reviewer for many highly respected journals and conferences, and is Associate Editor of *Journal of Global Information Management* (*JGIM*) and *Journal of Information Technology Research* (*JITR*). His research activity led to him receiving the Research Award "Scientific Computer Society of Spain and BBVA Foundation" as one of the best young researchers in Computer Science in Spain in 2019. As a software engineer and developer, he is involved in many different technical communities and open source initiatives.

Review

Contributions of Machine Learning Models towards Student Academic Performance Prediction: A Systematic Review

Prasanalakshmi Balaji [1,2,*], **Salem Alelyani** [1,3], **Ayman Qahmash** [1,3] **and Mohamed Mohana** [1]

[1] Center for Artificial Intelligence (CAI), King Khalid University, Abha 61421, Saudi Arabia; s.alelyani@kku.edu.sa (S.A.); a.qahmash@kku.edu.sa (A.Q.); mmuhanna@kku.edu.sa (M.M.)
[2] College of Science and Arts, King Khalid University, Abha 61421, Saudi Arabia
[3] College of Computer Science, King Khalid University, Abha 61421, Saudi Arabia
* Correspondence: prengaraj@kku.edu.sa

Abstract: Machine learning is emerging nowadays as an important tool for decision support in many areas of research. In the field of education, both educational organizations and students are the target beneficiaries. It facilitates the educational sector in predicting the student's outcome at the end of their course and for the students in deciding to choose a suitable course for them based on their performances in previous exams and other behavioral features. In this study, a systematic literature review is performed to extract the algorithms and the features that have been used in the prediction studies. Based on the search criteria, 2700 articles were initially considered. Using specified inclusion and exclusion criteria, quality scores were provided, and up to 56 articles were filtered for further analysis. The utmost care was taken in studying the features utilized, database used, algorithms implemented, and the future directions as recommended by researchers. The features were classified as demographic, academic, and behavioral features, and finally, only 34 articles with these features were finalized, whose details of study are provided. Based on the results obtained from the systematic review, we conclude that the machine learning techniques have the ability to predict the students' performance based on specified features as categorized and can be used by students as well as academic institutions. A specific machine learning model identification for the purpose of student academic performance prediction would not be feasible, since each paper taken for review involves different datasets and does not include benchmark datasets. However, the application of the machine learning techniques in educational mining is still limited, and a greater number of studies should be carried out in order to obtain well-formed and generalizable results. We provide future guidelines to practitioners and researchers based on the results obtained in this work.

Keywords: educational mining; machine learning; artificial intelligence; decision support systems; systematic literature review

Citation: Balaji, P.; Alelyani, S.; Qahmash, A.; Mohana, M. Contributions of Machine Learning Models towards Student Academic Performance Prediction: A Systematic Review. *Appl. Sci.* **2021**, *11*, 10007. https://doi.org/10.3390/app112110007

Academic Editor: Juan Cruz-Benito

Received: 7 August 2021
Accepted: 6 October 2021
Published: 26 October 2021

Publisher's Note: MDPI stays neutral with regard to jurisdictional claims in published maps and institutional affiliations.

1. Introduction

In recent centuries, the academic performances of students have been appraised on the basis of memory-related tests or regular examinations and by comparing their performances to identify the factors for predicting their academic excellence. In the contemporary world, there are full-fledged, developed, and advanced technologies that enable an individual from any domain, even with minimal programming knowledge, to predict their future data. Machine learning (ML) is now a prevalent technology to forecast data ranging from supermarkets to astronomical realms. Academicians and administrative personnel use data to predict a student's performance during the time of admission, predict the job scope for a student at the time of course completion or the dropout based on the aggregate numbers from the entire set of students, or gauge a particular student's success or failure rate in the subsequent grades. These have even led to recommendation systems for the students to select their area of expertise. These recommendation systems started its implementation from higher secondary schools [1], predicting the retention of students [2], family tutoring and recommender systems [3–6].With an enormous growth of research contributions in the field of big data and ML, learning analytics and supportive learning have also shown their growth in education. Education institutions have expressed their interest in predicting students' performance or related model development to estimate their own students' performance. This prediction is anticipated to favor the growth of their institutions. Such model developments are likely to support the creation of follow-up actions that may be taken to set up remedial actions on the drawbacks associated with student comprehension and to rectify them.

The objective of this systematic survey was to delineate completed, implemented, and published ideas of various researchers, starting from the earliest works to the most recent ones. Furthermore, this study aimed to understand the rate of success of the implemented ML-related models in specific domains of research to predict student academic performance. Even though there are a number of existing ML algorithms, only a few exist in every category based on the area of interest taken up for analysis. The regression algorithms stand to prove their accuracy in the prediction of student academic performance. The regression algorithms [6–11] stand by classification algorithms [12–17] to enhance their prediction accuracy by means of ensemble methods. As analyzed, this survey starts with developing an systematic literature review (SLR) model, which provides pertinent ideas to novice researchers on the algorithms used or articles published, and their results obtained in the domain of predicting student academic performance. This may potentially lead to the creation of a new ML model that can yield much higher accuracy with limited usage of resources. The purpose of this SLR is to summarize and clarify the available and accessible resources of the previously published articles. The rest of this paper is organized as follows: Section 2 discusses the methodology adopted, followed by Section 3 that summarizes the findings, followed by an elaborate discussion of the same. Section 4 outlines the implications of this SLR, the limitations of this study, and finally, suggestions for prospective future research.

2. Review Methodology

The purpose of this SLR is to study the published articles in the domain of student academic performance prediction with the help of machine learning (ML) or artificial intelligence (AI)-related models. To acquire a deep insight of the previous works published, the domain of interest was analyzed from multifarious dimensions. To perform this SLR in a well-formed structure, the methodology underwent five different stages as shown in Figure 1.

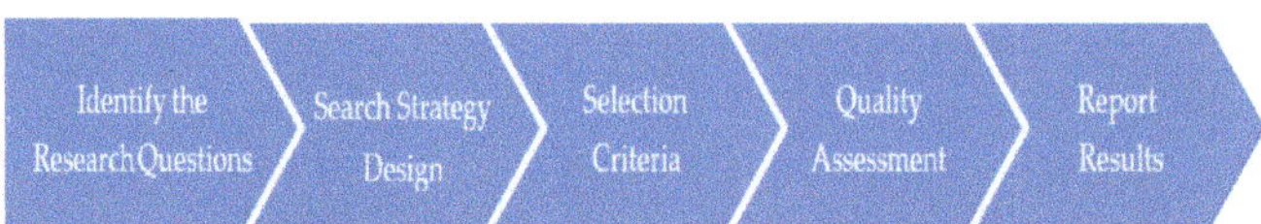

Figure 1. Implemented systematic literature review protocol.

The first step initiated with the identification of the research questions, which provided clear data on the nature of publications presented so far in the specified area of research. This identification of research questions, in turn, provided a coherent picture of the design of search strategy. The results obtained from the defined search strategy narrowed down this study to precisely define the selection criteria to filter the articles that were pertinent to the real necessity of this study. To filter even further based on the "quality," a scoring system related to the testing of the quality of the selected articles was framed. The final corpus of articles was evaluated, and the results are reported in this paper.

2.1. Research Question (RQ) Identification

The framed SLR aimed to provide and assess the empirical evidence from the studies that deployed ML or AI models in predicting the student academic performance. The motivation behind developing these RQs relied on the real focus of this systematic review. The initial process of SLR and the perfect base to perform the SLR was formed with the exact definition of the RQs. Five principal RQs were framed to explicate the exact idea of this systematic review.

The research questions were framed such that the articles responding either partially or perfectly alone stands in the filter of articles. These sustained articles proceeded further for evaluation to describe the concepts of machine learning application in the field of educational mining.

RQ1: What are the different ML models/techniques used for student academic performance prediction?

The aim of this research question is to understand the models that have been implemented for predicting student academic performance. The models/techniques used are analyzed to obtain an insight of the most frequently used methods, new proposed methods, and methods that provide better results or performance metrics. The reader of the article will be able to find a list of such methods that are utilized and proved by various researchers so that the budding researchers can adopt new ideas from existing works.

RQ2: What are the various estimation methods/metrics used? What is the performance measure used to appraise the performance of the models in the described problem area?

This research question has been framed with the target to identify the metrics that have been used to measure the precision of the developed model. Furthermore, it aims to assess the way the referred articles speak of their credibility and accuracy in proving the purpose of the developed model. Even though the metrics used for analyzing the machine learning models are standardized, the values obtained by various methods on different databases speak to the importance of each feature and its contribution towards the performance metrics discussed in the relevant articles.

RQ3: Are there any datasets and collection methods; if they exist, is their usage specified?

RQ3 responds with the quality of analysis made, so that the size of datasets reveals their proportion of reliability. The datasets used in the referred articles are considered as a research question to show the features taken under consideration and the importance level of each feature towards arriving at the best model and better performance measure.

RQ4: Are there any guidelines on the number of features considered or the features used?

This research question aims at finding the data collected and the source and identifying the effective features in the dataset. The guidelines regarding the features used provide an idea during the feature extraction process of developing an ML-based model or a prototype. These guidelines discussed in different articles of multiple features show the insights that each author has attained during their research. The readers will be able to identify the importance of choosing the features or the justification provided by the authors in eliminating the consideration of certain features.

RQ5: Are enough comparisons made to prove the reliability of the proposed model?

The models that are proposed in the cluster of articles are to be segregated and selected for a further examination based on the comparative measures that were taken to validate the proposed works as adequate and substantial and that they surpass the previously presented works in the pertinent literature. Even though every model when proposed seems to prove its innovation, solid proof is needed to say that the proposed model is genuine. Hence, a considerable number of methods that were existing should be taken into consideration and analyzed for improvement in the performance measure of the proposed system. Hence, a bird's-eye definition is needed in each article to prove its contribution.

2.2. Search Strategy

AI expanded its level of implementation combined with data mining and knowledge discovery into a notable field of model development in the form of ML, which grew further into another level of deep learning (DL). This paper predominantly focuses on ML and AI as a peripheral aspect of implementation to develop a model in the problem of predicting student academic performance.

2.2.1. Search Strategy Design

To narrow down the search among thousands of published research articles, the search queries (SQs) were defined clearly and delimited by the refined queries. The input terms involve "machine learning", "artificial intelligence," "academic performance prediction," "student academic performance," and "student success prediction." Even though the search could have been performed in all fields of metadata, it was restricted to Title, Abstract, and Keywords. The corpus for the synthesis was created through a metadata search on the article indexed in six major libraries of academic publications, namely Google Scholar, Web of Science (WoS), Scopus, ScienceDirect, SpringerLink, and IEEE Explore. The syntax of the search could be altered based on the database requirements. The search period of the database varied from 1959 to 2020 and the articles that are yet to be published, and based on the period of application, this may have been reduced further. Only the latest articles in this application were considered.

The SQs were used in the following formats and varied based on the database conditions of search.

SQ1: ["machine learning" AND "academic performance prediction"] (("title" AND "abstract") OR "keywords")

SQ2: [(("machine learning" OR "artificial intelligence") AND ("academic performance prediction" OR "student academic performance"))] (("title" AND "abstract") OR "keywords")

SQ3: ["machine learning" AND "student success prediction"] (("title" AND "abstract") OR "keywords")

The input search query, SQ1–SQ3, gave a generalized format; however, each database had its own form of SQ. Thereby, the search was performed accordingly.

2.2.2. Selection Criteria

The entire procedure of selection criteria was divided into two phases. Phase 1, termed as the collection and analysis phase, comprised article collection, removing duplicates, and applying inclusion and exclusion criteria, as shown in Figure 2. Phase 2, termed as the synthesis phase, forwarded the refined articles to proceed further with quality analysis to refine further and narrow down the articles to analyze the methods and find results for the defined five research questions.

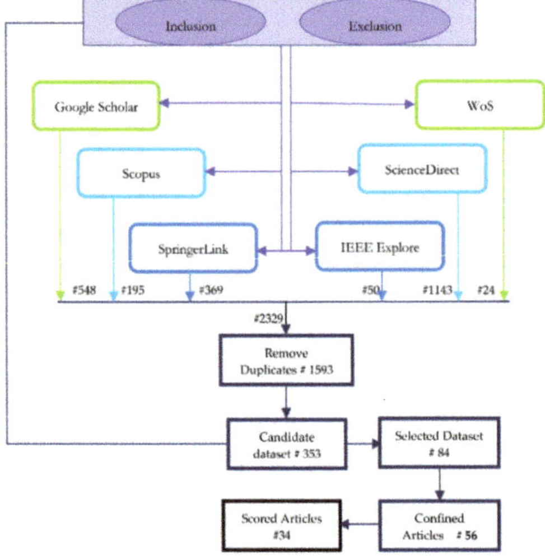

Figure 2. Phase 1: collection and analysis.

The set of articles collected as a corpus based on the SQs (SQ1–SQ3), counting to 2239 articles, included some of the inclusion and exclusion criteria except that could not be applied at this stage. The collected articles were refined further to remove duplicates and proceed with the reapplication of missed out inclusion or exclusion criteria, if any.

Inclusion criteria:

- Using ML to analyze the academic performance;
- Using ML to preprocess the modeling data;
- Comparative assessment of various ML methods and their results obtained;
- Journal versions—for duplicate articles, the recently published article is considered;
- Articles including online education assessment;
- Academic performance and recommendation systems.

Exclusion Criteria:

- Review articles;
- Book chapters;
- Factor analysis;
- Articles not written in the English language.

Step 1: Extract all the articles from the six data sources with the predefined criteria for inclusion and exclusion.

Using the four SQs as defined, a corpus of articles was amassed. A total of 2329 articles were collected based on the defined SQs as shown in Table 1. From this basic search retrieval, the duplicate articles were removed.

Table 1. A Summary of the collected articles.

SQ	SQ1	SQ2	SQ3	Database-Wise Count
Google Scholar	70	465	13	548
WoS	171	24	0	195
Scopus	3	170	1	369
ScienceDirect	8	41	1	50
SpringerLink	1043	0	100	1143
IEEE Explore	0	24	0	24
Query-Wise Count	1295	724	115	2329

Step 2: Remove duplicates.

Duplicates that are obtained using search query. Since the articles chosen might find citations in several sources, this step was carried out. After the elimination of the duplicates, a total of 1593 articles for further processing were selected. For duplication of the article title from multiple databases, the database that had the article in its creamy layer, the top 50% of the total article is retained and the article in other citing database sources is eliminated.

Step 3: Re-apply inclusion and exclusion criteria if needed.

After the application of the same to the Title, Abstract, and Keywords, the resultant set comprised 353 articles.

Step 4: Manual refinement of corpus.

After manual refinement by analyzing each title, the final obtained result set mentioned as the selected dataset was 80 articles. Firstly, manual refinement was carried out by eliminating the articles that had a similar combination of title and method that seemed to be repeated. Secondly, in some cases, even if there was a match of keywords applied in the search query abstract of some articles, it did not reflect the necessary information of the review considered. The main source of manual refinement is the abstract of the article. Only the recent and complete study articles were considered.

Step 5: The selected articles were then evaluated using the quality assessment as discussed in the next section to finalize with 56 articles.

2.3. Study Quality Assessment

After refinement with consideration of these basic quality measures, the articles were analyzed for quality assessment as enumerated in Figure 3. The selected dataset reduced to the most suitable 56 articles, pertinent to the requirement of research for further investigation.

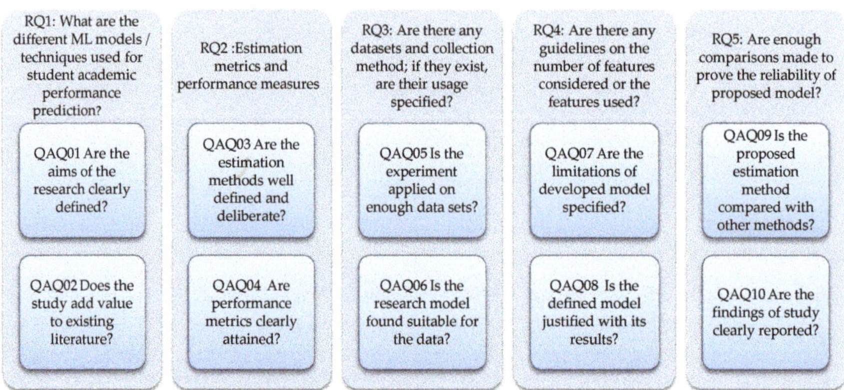

Figure 3. Quality assessment questions and related research questions.

3. Results and Discussions

A fundamental eligibility criterion for selecting the articles was that they could answer the research questions framed. Table 2 provides evidence corroborating the selection of articles for study. Based on the questions each article was able to answer and the data that could be obtained for further quality score assessment of the article, the summary of articles taken up for study is tabulated. These tabulating aspects preceded in the rest of the article reveals the entire systematic literature review of the article. The pictorial and tabular depiction of results aims to give a better insight on the study carried over.

Table 2. Summary of selected studies.

Ref.	RQ1	RQ2	RQ3	RQ4	RQ5	Ref.	RQ1	RQ2	RQ3	RQ4	RQ5	Ref.	RQ1	RQ2	RQ3	RQ4	RQ5	Ref.	RQ1	RQ2	RQ3	RQ4	RQ5
[1]	*	-	*	-	-	[2]	*	-	-	-	-	[3]	*	*	*	-	*	[4]	*	-	-	-	-
[5]	*	*	*	-	-	[6]	*	*	*	*	*	[7]	*	-	-	-	-	[8]	*	*	*	*	*
[9]	*	*	*	-	-	[10]	*	*	*	*	*	[11]	*	*	*	-	-	[12]	*	*	*	*	*
[13]	*	*	*	*	*	[14]	*	*	*	-	*	[15]	*	-	-	-	-	[16]	*	*	*	*	*
[17]	*	*	-	*	*	[18]	*	*	*	*	*	[19]	*	*	*	*	*	[20]	*	-	-	-	-
[21]	*	*	*	-	*	[22]	*	-	-	-	*	[23]	*	*	*	*	*	[24]	*	*	*	*	*
[25]	*	*	*	*	*	[26]	*	*	*	*	-	[27]	*	-	-	-	*	[28]	*	*	*	*	-
[29]	*	-	-	-	-	[30]	*	-	-	-	*	[31]	*	*	*	*	*	[32]	*	*	*	*	*
[33]	*	*	*	-	*	[34]	*	*	-	-	*	[35]	*	*	*	*	*	[36]	*	*	*	*	*
[37]	*	*	*	*	*	[38]	*	*	-	-	*	[39]	*	*	*	*	*	[40]	*	*	*	*	-
[41]	*	*	-	-	-	[42]	*	*	*	-	*	[43]	*	*	-	*	*	[44]	*	*	*	-	*
[45]	*	*	*	*	*	[46]	*	*	*	*	*	[47]	*	*	*	*	*	[48]	*	*	*	*	*
[49]	*	*	*	-	-	[50]	*	*	*	-	-	[51]	*	-	-	-	-	[52]	*	*	*	*	*
[53]	*	-	-	-	-	[54]	*	*	*	-	*	[55]	*	*	*	*	*	[56]	*	*	*	-	*
[57]	*	*	*	-	-	[58]	*	*	*	-	-	[59]	*	*	*	*	*	[60]	*	*	*	-	*
[61]	*	*	*	*	*	[62]	*	*	*	*	*	[63]	*	*	*	-	*	[64]	*	*	*	*	*
[65]	*	*	-	*	*	[66]	*	*	*	*	*	[67]	*	*	*	*	*	[68]	*	*	*	-	-
[69]	*	*	*	*	*	[70]	*	-	-	-	-	[71]	*	-	*	-	-	[72]	*	*	-	*	*
[73]	*	-	*	-	-	[74]	*	*	*	*	*	[75]	*	*	-	*	*	[76]	*	*	-	*	*
[77]	*	*	*	*	*	[78]	*	*	-	*	*	[79]	*	*	*	*	*	[80]	*	*	*	*	*

Note: * represents that the research article responds to one or both of the QAQ in each RQ, others do not.

3.1. Overview of the Selected Studies

As the studies extracted from SpringerLink and ScienceDirect in the candidate datasets consisted of more research databanks, the search source was restricted to four of the indexing sources in majority, namely Scopus, WoS, IEEE Explore, and with the least contribution from Google Scholar. A total of 56.6% of the selected papers were taken from Scopus and the next 38.5% was contributed by the articles taken from WoS for this study. The remaining 5% of the articles were found interesting from IEEE Explore and Google Scholar. The details are depicted in Table 3.

Table 3. Distribution of the selected studies.

Indexing Source	# of Articles	%
Scopus	47	56.6
WoS	32	38.5
IEEE Explore	3	3.6
Google Scholar	1	1.2

As shown in Figure 3, the research questions were extended to 10 quality assessment questions based on which the scores of Table 4 are calculated.

Table 4. Quality scores of selected studies.

Ref.	QAQ01	QAQ02	QAQ03	QAQ04	QAQ05	QAQ06	QAQ07	QAQ08	QAQ09	QAQ10	SCORE
[1]	1	1	0	0	1	1	0	0	0	0	4
[3]	1	1	1	1	1	1	0	0	1	1	8
[5]	1	1	1	1	1	1	0	0	0	0	6
[7]	1	1	1	1	1	1	0	0	1	1	8
[9]	1	1	1	1	1	1	1	0	1	1	9
[11]	1	1	1	1	0	0	0	0	1	1	6
[13]	1	1	1	1	0	1	0	0	0	0	5
[15]	1	1	1	1	0	0	1	1	1	1	8
[17]	1	1	1	1	1	1	1	1	1	1	10
[19]	1	1	1	1	1	1	1	1	1	1	10
[21]	1	1	1	1	1	1	0	0	0	0	6
[23]	1	1	1	1	1	1	1	1	1	1	10
[25]	1	1	0	0	0	0	0	0	0	0	2
[27]	1	1	1	1	1	0	1	1	1	1	9
[29]	1	1	1	1	1	1	1	1	1	1	10
[31]	1	1	1	1	1	0	0	0	1	1	8
[33]	1	1	1	1	0	1	1	1	1	1	9
[35]	1	1	1	1	0	1	1	0	1	1	8
[37]	1	1	1	1	0	1	1	1	1	1	9
[39]	1	1	1	1	1	1	0	0	1	1	8
[41]	1	1	1	1	1	1	1	1	1	1	10
[43]	1	1	1	1	0	0	0	0	1	1	6
[45]	1	1	1	1	1	1	1	1	1	1	10
[47]	1	1	1	1	0	1	0	0	1	1	7
[49]	1	1	1	1	0	1	0	0	1	1	7
[51]	1	1	1	1	0	1	1	1	1	1	9
[53]	1	1	1	1	0	1	0	0	0	0	5
[55]	1	1	0	0	0	0	0	0	0	0	2

Ref.	QAQ01	QAQ02	QAQ03	QAQ04	QAQ05	QAQ06	QAQ07	QAQ08	QAQ09	QAQ10	SCORE
[2]	1	1	0	0	0	0	0	0	0	0	2
[4]	1	1	0	0	0	0	0	0	0	0	2
[6]	1	1	1	1	1	1	1	1	1	1	10
[8]	1	1	1	1	1	1	0	0	0	0	6
[10]	1	1	1	1	1	1	1	1	1	1	10
[12]	1	1	1	1	0	1	1	1	1	1	9
[14]	1	1	1	1	0	1	1	1	1	1	9
[16]	1	1	1	1	0	0	1	1	1	1	8
[18]	1	1	0	0	0	0	0	0	0	0	2
[20]	1	1	1	1	0	1	1	1	1	1	9
[22]	1	1	1	1	1	1	1	1	1	1	10
[24]	1	1	1	1	1	1	0	0	1	1	8
[26]	1	1	1	1	1	1	1	1	1	1	10
[28]	1	1	1	1	1	1	1	1	1	1	10
[30]	1	1	0	0	0	0	0	0	0	0	2
[32]	1	1	0	0	0	0	0	0	0	0	2
[34]	1	1	1	1	1	1	1	1	1	1	10
[36]	1	1	0	0	0	0	0	0	0	0	2
[38]	1	1	1	1	0	1	1	0	1	1	8
[40]	1	1	1	1	1	1	1	0	1	1	9
[42]	1	1	1	1	0	1	1	1	1	1	9
[44]	1	1	1	1	1	1	1	1	1	1	10
[46]	1	1	1	1	1	0	0	0	0	0	5
[48]	1	1	1	1	0	0	1	1	1	1	8
[50]	1	1	1	1	0	1	1	0	1	1	8
[52]	1	1	1	1	0	1	1	1	1	1	9
[54]	1	1	1	1	1	1	0	0	0	0	6
[56]	1	1	1	1	1	1	1	1	1	1	10

Table 4. *Cont.*

Ref.	QAQ01	QAQ02	QAQ03	QAQ04	QAQ05	QAQ06	QAQ07	QAQ08	QAQ09	QAQ10	SCORE	Ref.	QAQ01	QAQ02	QAQ03	QAQ04	QAQ05	QAQ06	QAQ07	QAQ08	QAQ09	QAQ10	SCORE
[57]	1	1	0	0	0	0	0	0	0	0	2	[58]	1	1	1	1	0	1	0	0	1	1	7
[59]	1	1	1	1	0	1	1	0	1	1	8	[60]	1	1	1	1	0	1	0	0	1	1	7
[61]	1	1	1	1	1	1	0	0	0	0	6	[62]	1	1	1	1	1	1	1	0	1	1	9
[63]	1	1	1	1	0	1	0	0	1	1	7	[64]	1	1	1	1	1	1	1	0	1	1	9
[65]	1	1	1	1	1	1	1	1	1	1	10	[66]	1	1	1	1	1	1	1	0	1	1	9
[67]	1	1	1	1	0	1	0	0	1	1	7	[68]	1	1	1	1	1	1	1	1	1	1	10
[69]	1	1	1	1	1	0	1	1	1	1	9	[70]	1	1	1	1	1	1	1	1	1	1	10
[71]	1	1	1	1	0	1	0	0	0	0	5	[72]	1	1	1	1	1	1	1	1	1	1	10
[73]	1	1	0	0	0	0	0	0	0	0	2	[74]	1	1	0	0	0	1	0	0	0	0	3
[75]	1	1	1	1	0	0	1	1	1	1	8	[76]	1	1	0	0	1	1	0	0	0	0	4
[77]	1	1	1	1	1	1	1	1	1	1	10	[78]	1	1	1	1	0	0	1	1	1	1	8
[79]	1	1	1	1	0	1	1	1	1	1	9	[80]	1	1	1	1	1	1	1	0	1	1	9

The selected articles were taken into consideration to undergo quality score assessment. The scores were given on a scale of 1, 0.5, and 0 measuring a positive, partial, and negative response, respectively, to the quality assessment questionnaire that comprised 10 questions contributing to a total score of 10 for each selected article. Quality assessment attempts to weigh the studies and their importance to this survey. The scores were categorized as very high (9–10), high (7–8), medium (5–6), low (3–4), and very low (0–2). Each study under consideration could have a maximum score of 10 and a minimum score of 0.

Hence, the 80 studies taken into consideration as shown in Table 2 were reduced to 56 final articles for a further analysis. The quality assessment questionnaire was prepared such that the answers were derived in relevance to the research questions.

Table 5 shows the article distribution from the selected articles based on the quality score.

Table 5. Article distribution based on quality score.

Criteria	# of Articles	% of Articles
very high (9 ≤ score ≤ 10)	36	45%
high (7 ≤ score ≤ 8)	20	25%
medium (5 ≤ score ≤ 6)	11	13.75%
low (3 ≤ score ≤ 4)	3	3.75%
very low (score ≤ 2)	10	12.5%

3.2. Models and Metrics Used

The selected publications illustrate the reference and the ML methods to furnish an insight of the overview on the models developed and to provide an answer to RQ1. For ease of analysis, the ML algorithms branched are categorized under major classes. RQ1 addresses the ML models used. The entire set of ML models used in different articles is broadly categorized as decision trees (DT), neural network (NN), support vector machine (SVM), and ensemble method. RQ1 was supported by the graphical data in Figure 4, depicting the ML methods used in the selected articles of study and the frequency of their use. Figure 4 shows that 32% of the used models contribute to ensemble models, 22% to neural network models, 26% of decision tree models, and 14% of other ML algorithms.

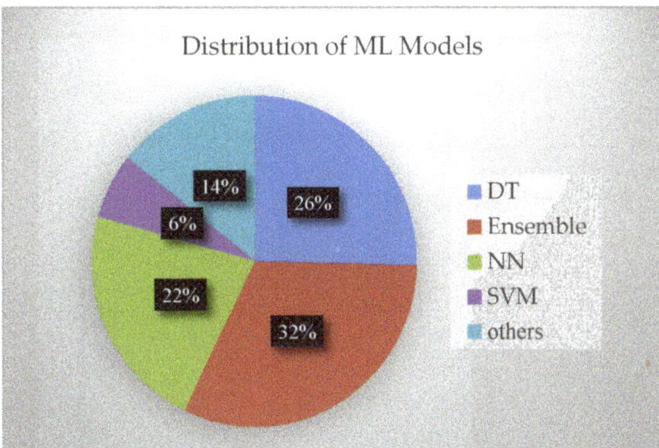

Figure 4. Graphical illustration of ML methods used.

As defined by Patrick [81], the taxonomy of algorithms that were utilized for the purpose of study are shown in Figure 5a,b. The taxonomy defined is based on the mathematical impact of the algorithms used. The following subsections gives a brief notation of the several mathematical idea-based models as given by Patrick [81].

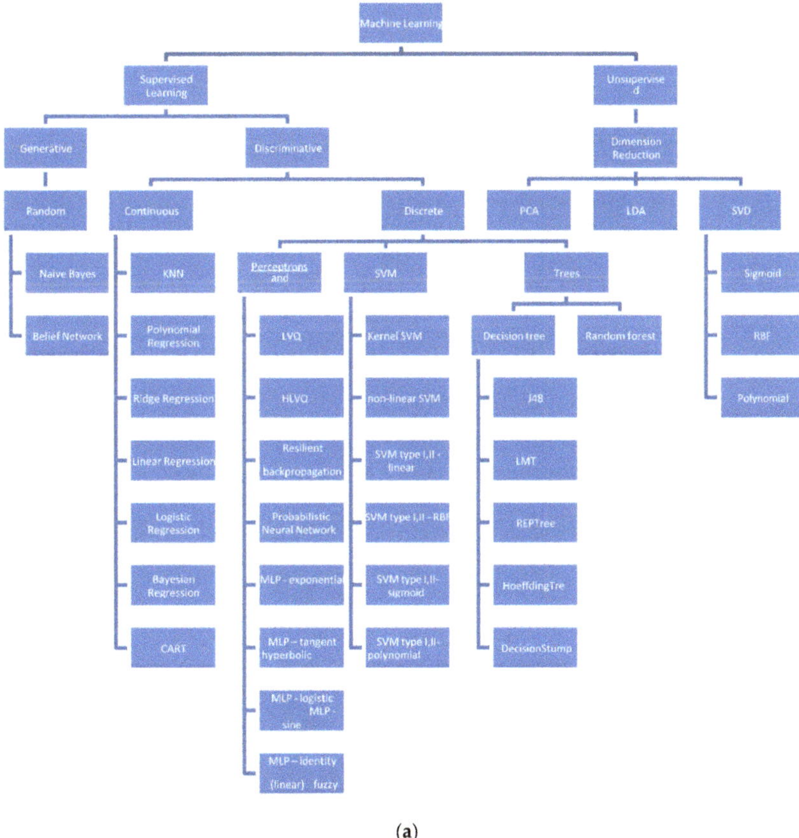

(**a**)

Figure 5. *Cont.*

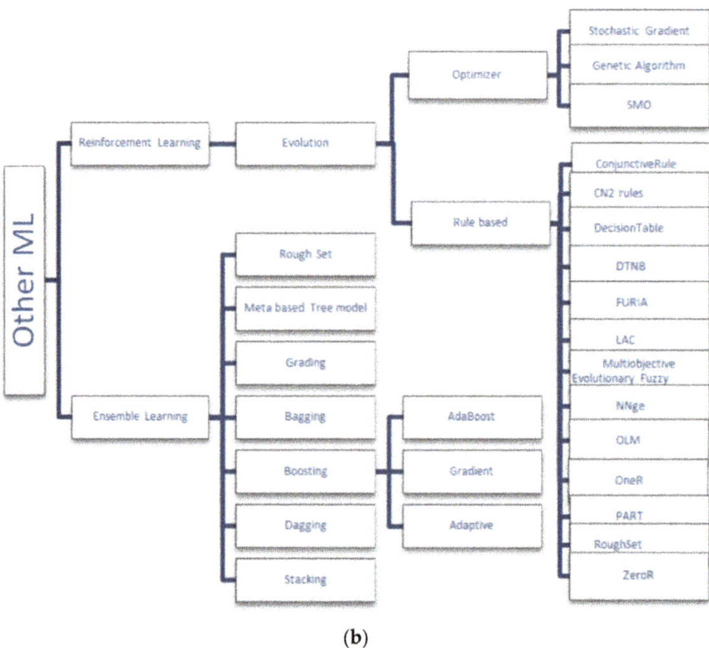

(b)

Figure 5. (**a**) Taxonomy of the algorithms used. (**b**) Taxonomy of other machine learning models used.

3.2.1. Supervised Learning Algorithms

Given a set of data points $\{x_1, \ldots, x_m\}$ associated to a set of outcomes $\{y_1, \ldots, y_m\}$, the aim of the supervised learning algorithms is to build a classifier that can predict y from x. The supervised learning algorithm prediction can be a regression model producing a continuous output or a classification model predicting a class of the given input values. A broad classification of supervised learning involves two types of mathematical concepts to perform either classification of regression model formulation. Logistic regression, support vector machine, and conditional random fields are popular discriminative models; naïve Bayes, Bayesian networks, and hidden Markov models are commonly used generative models. The supervised model is branched up as generative and discriminative models as in Figure 6. The generative model learns the probability distributions of the data and estimates the conditional probability $P(x|y)$ to then deduce to the posterior $P(y|x)$, whereas the discriminative model creates a decision boundary to directly estimate $P(y|x)$.

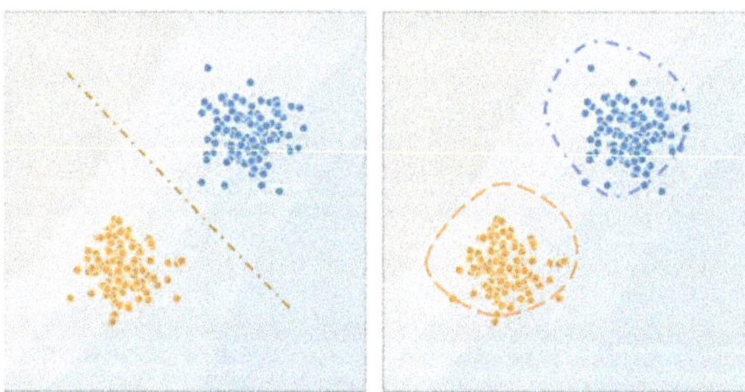

Figure 6. Supervised learning models.

In a generative random model, the two models that were used by different authors include naïve Bayes and belief networks. Naïve Bayes assumes that all features are independent, whereas belief networks allows the user to specify which attributes are, in fact, conditionally independent.

In supervised learning, the hypothesis is noted as h_θ, the model that we choose. For a given input data x_i, the model prediction output is $h_\theta(x_i)$. A loss function is given as $L{:}(z,y) \in \mathbb{R} \times Y \longmapsto L(z,y) \in \mathbb{R}$ that takes as inputs the predicted value z corresponding to the real data value y and outputs how different they are. Some of the loss functions are least squared error, logistic loss, hinge loss, and cross entropy. The loss function then contributes to the calculation of cost function as $J(\theta) = \sum_{i=1}^{m} L(h_\theta(x_i), y_i)$. The update rule for gradient descent is expressed in terms of the cost function calculated, and the learning rate $\alpha \in \mathbb{R}$ is given as $\theta \leftarrow \theta - \alpha \nabla J(\theta)$. With the known parameters $L(\theta)$, the likelihood and the parameters θ, the optimal parameters are determined as $\theta_{optimal} = \text{argmax}(L(\theta))$.

Some of the linear discriminative models included in the survey articles are linear regression, logistic regression, polynomial regression, ridge regression, and the non-parametric discriminative model, which includes K-nearest neighbor. While discrete discriminative models include support vector machine models, neural networks, and trees in several variants. However, the linear discriminative algorithm works in its own fashion, and the articles taken up for study have included some of the mentionable variations in their work. They include the Widrow–Hoff rule and locally weighted regression parameters in the calculation of optimal parameters. Support vector machines have used the concepts of Lagragian multipliers and Kernel and optimal classifiers in different notations.

3.2.2. Unsupervised Learning Algorithms

Unsupervised learning algorithm takes into account the aim of finding hidden patterns from the input data, provided output labels do not exist. The major concentrations of the authors were found in clustering, Jensen's inequality, mixture of Gaussians, and expectation maximization. Most of the articles on unsupervised learning tried to attain their pattern of clustering by finding patterns using dimension reduction techniques. These dimension reduction techniques find the variance maximizing directions onto which to project the data. Some of the metrics used to evaluate the clustering are the Davies–Bouldin index, popularly known as DB index, which calculates the average distance of all points in a particular cluster from the cluster centroid, and the Dunn index that calculates the ratio between the minimum inter-cluster distance to the maximum intra-cluster distance. The Dunn index showed an increase as the performance of the model improved.

Many ensemble learning and reinforcement learning algorithms were taken into account in the survey made. Even though all the articles taken for study could not be illustrated, some of the models are given in brief note. For better understanding of the models used, they are best illustrated in Figure 5a,b.

The articles that contribute to the categorization are shown in Tables A1–A4 of the Appendix A. Based on the quality score, the articles are segregated. Even though there exist some discrepancies with the presence of metrics in the articles, they were also considered for quality assessment, where they obtained a minimal score. The evaluation metrics or the principles used were not found in most of the articles, which could not provide a proper insight on the data or the model used.

The entire set of the selected articles were evaluated and analyzed with respect to the performance metrics, which gives a response to RQ2. The set of algorithms, as discussed in the previous section, takes up the major category based on the way those algorithms function as classification and regression, and in some articles, an ensemble of classification and regression algorithms were used. The usage of articles in these categories is analyzed as shown in Figure 7.

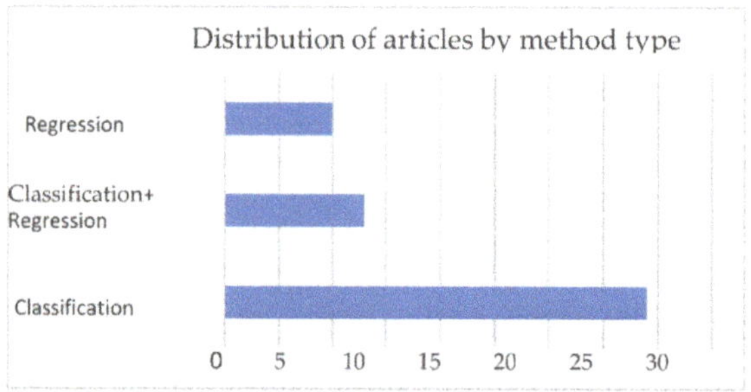

Figure 7. Distribution of ML models used.

A classification problem is the one where the dependent variable is a categorical one. Classification models entail algorithms such as logistic regression, DT, random forest, and naïve Bayes. Model performance metrics are estimated based on the values obtained from confusion matrix, accuracy score, classification report, receiver operating characteristic (RoC), and area under curve (AUC), confusion matrix being an intuitive metrics to determine the accuracy of the given model is suitable for a multiclass classification problem. The performance metrics with respect to the classification algorithms taken up for study are listed in Table 6.

Table 6. Performance metrics—classification algorithms.

Performance Metrics	Measures	SVM	DT	NN	Ensemble
Accuracy	Count	4	15	14	16
	Mean	60	80	86.9	84.7
	Min	93.8	79	51.9	61
	Max	95	98.94	100	98.5
	Std. Dev	40.86	24.46	12.16	24.08
Sensitivity/Recall	Count	1	9	7	12
	Mean	89	88.13	83.2	71.8
	Min	89	75	51.9	48
	Max	89	100	100	96.3
	Std. Dev	-	9.47	17.8	34.5

Performance Metrics	Measures	SVM	DT	NN	Ensemble
F1	Count	1	8	6	13
	Mean	89	82.5	81.65	72.7
	Min	89	71.9	49.4	53
	Max	89	95.6	100	98.2
	Std. Dev	-	9.2	17.08	33.9
AUC	Count	1	5	3	3
	Mean	75	89.2	84	89.7
	Min	75	68	63.5	77.7
	Max	75	99.4	100	99.6
	Std. Dev	-	12	18.6	11.1

Performance Metrics	Measures	SVM	DT	NN	Ensemble
Precision	Count	1	10	7	11
	Mean	89	85.9	87.87	75.1
	Min	89	70.3	48.6	95
	Max	89	98.3	100	97
	Std. Dev	-	8.77	17.8	28.8
Specificity	Count	2	-	5	1
	Mean	47.28	-	73	100
	Min	0.97	-	54.6	100
	Max	94	-	85.16	100
	Std. Dev	65	-	14.73	-

Confusion matrix depicts the overall performance of the model, and accuracy reveals the number of correct predictions made by the model.

Regression problems are the ones wherein we find a linear relationship between the target variables and the predictors. In such problems, the target variable holds a continuous value. Such methods are typically used for forecasting. Regression models include algorithms such as linear regression, DT, random forest, and SVM.

The performance metrics of the regression problems are identified as mean absolute error (MAE), mean square error (MSE), root mean square error (RMSE), and R-squared error. An MAE value of 0 indicates no error or perfect predictions. An MSE estimated to zero means that the estimator predicts observations of the parameter with perfect accuracy. Root mean squared error (RMSE) measures the average magnitude of the error by taking the square root of the average of squared differences between the predicted and actual observation. The RMSE will always be larger than or equal to the MAE; the greater the difference between them, the greater the variance in the individual errors in the sample. If RMSE = MAE, then all the errors are of the same magnitude. R-squared score is the proportion of the variance in the dependent variable that is predictable from the independent variable(s). It is known as the coefficient of determination. The value of R^2 lies between 0 and 1, where 0 means no fit and 1 implies a perfect fit. The performance metrics with respect to the regression algorithms taken up for study are listed in Table 7. Research question analysis on the corpus yielded a valuable report on the datasets utilized for the analysis of the proposed models. Most of the authors have performed the analysis on collected real-time data from various educational institutions.

Table 7. Performance metrics—regression algorithms.

	Measures	SVM	DT	NN	ENSEMBLE
MSE	Count	-	1.00	1.00	2.00
MSE	Mean	-	0.05	75.90	7.60
MSE	Min	-	0.05	0.75	0.14
MSE	Max	-	0.05	0.75	0.15
MSE	Std. Dev	-	-	-	10.59
MAE	Count	-	-	-	3.00
MAE	Mean	-	-	-	9.60
MAE	Min	-	-	-	55.00
MAE	Max	-	-	-	12.15
MAE	Std. Dev	-	-	-	0.35

	Measures	SVM	DT	NN	ENSEMBLE
RMSE	Count	-	2.00	-	4.00
RMSE	Mean	-	0.35	-	6.14
RMSE	Min	-	0.21	-	56.00
RMSE	Max	-	0.50	-	17.90
RMSE	Std. Dev	-	0.21	-	0.79
Error	Count	-	1.00	2.00	3.00
Error	Mean	-	6.96	77.50	15.17
Error	Min	-	6.96	0.30	12.50
Error	Max	-	6.96	15.20	18.30
Error	Std. Dev	-	-	10.50	2.95

However, only a few articles reflect their testing and validation on open data sources. The box and whisker plots in Figure 8 denote the percentile of the values obtained against each performance metric. Even though there exists certain outliers for a few performance metrics, they can be overlooked. Table 8 displays the specific articles that have utilized the mentioned performance metrics and the number of articles that have utilized these measures.

Table 8. Evaluation parameters used.

Performance Metric	References	#
Accuracy	[3,6,8,11,12,14–18,20,22,24–27,29,31,34–40], [42–47,51,53,54,56,57,59,60,63–67,69,72–74,76–79]	51
Sensitivity	[16,25,31,43,53,56,64]	7
Specificity	[3,16,25,43,53,56,69,72,77]	9
AUC	[12,17,19,22,29,35,38,41,46,55,58,61,80]	13
Recall	[6,7,11,12,15–18,22,24,27,29,32,35,37] [39,41,45,46,55,59–61,63,64,66,76,78,80]	29
Precision	[7,11,12,15,17,18,22,24,25,27,29,32,34,35,37] [39,41,45,46,55–57,59,60,63,64,66,76,78,80]	30
F1	[6,7,11,12,15–17,22,24,27,29,32,35] [37–39,43,46,55,59,61,63,64,66,76,78,80]	27
MSE	[6,26,43,44]	4
MAE	[30,33,42,63,75]	5
RMSE	[7,10,30,54,55,61,74]	7
Error	[18,45,55,56,58,61,80]	7

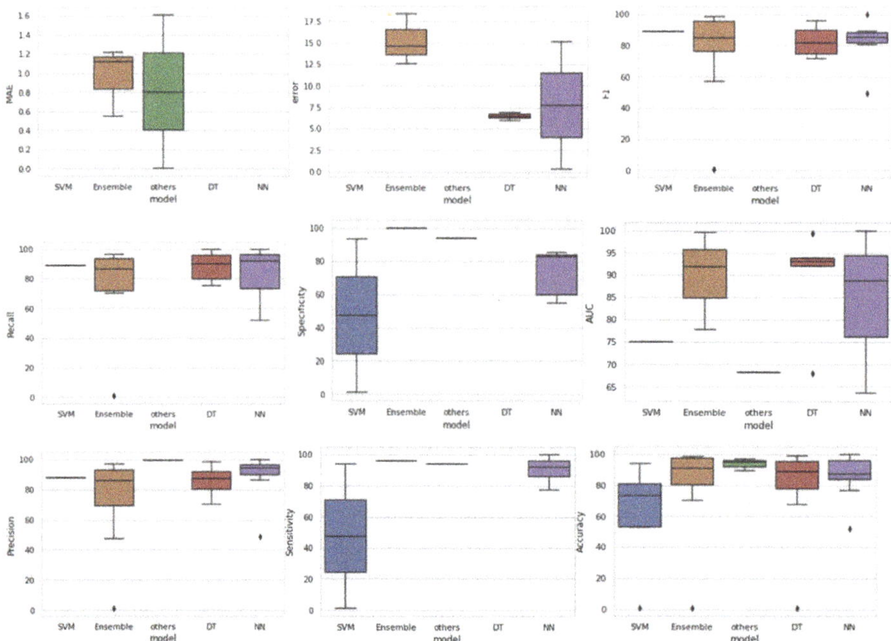

Figure 8. Box and whisker plots—performance metrics.

3.3. Dataset Preparation and Utilization

RQ3 enquired about the datasets, their collection methods, and their details of usage. RQ3 is provided with a score in quality assessment based on the number of datasets used. Considering the datasets used in the selected corpus of articles and their maximum and minimum sizes, the score varied from 100s, 1000s, and 10,000s as 0, 0.5, and 1, respectively. Since minimum data size could not prove the credibility of results obtained by the author's experiment, they were provided with the least evaluation score. Approximately 75% of the articles were assessed for their dataset with the ML algorithms and reported the performance metrics based on the self-collected data from their own source. The remaining 25% of the articles used existing datasets of academic performances. The related articles' references are specified in Table 8. The datasets used are mentioned in the Tables A1–A4 of the Appendix A. The datasets are not detailed in this section, since they are not benchmark datasets, but the parameters that were considered by various datasets collected are consolidated in Table 9.

3.4. Feature Description and Usage

Even though RQs 4 and 5 speak of the feature mentions and their usage in the research articles under study, they play a central role in deciding the features to be concentrated and the way data collection can take place in future in order to proceed with effective research on academic performance prediction.

Some of the features mentioned as a group in the research articles are social, demographic, personal, academic, extracurricular, and previous academic record. Even though the categorization was made in this common aspect, the individual components contributing to the research conducted varied in accordance with the educational institution, their geography, and their previous experience with students. The data do not only limit themselves to category of education institution or mode of study; it varies from online education university, online education courses, regular academic universities, colleges, schools, and others.

The category of features to their accuracy and their importance is shown in Figures 9 and 10. Figure 9 includes the details of the importance given to each feature category. Out of the scored articles, around 39% of articles contributed with three-feature set importance, dual-feature set importance is given by 30% of articles, and one-feature set is given importance by 32% of articles. Additionally, it is noted that minimal accuracy stays when considering only the behavioral features. However, the accuracy stays equal with demographic–academic dual feature and with academic features contributing to an average accuracy when all the three features are considered.

Based on the features described in the articles, they are broadly classified based on demographic, academic, and behavioral features. Out of the selected 56 articles of study, only 34 articles sustained delivering their features and their nature of importance. Table 9 specifies the detailed list of parameters that have been adopted in the studies. The quality assessment score is contributed as a response received from RQ 6 as shown in Table 4. The number of comparisons made in the proposed model to prove that its excellence against other models was considered. The taxonomy broadly explains the models used in each category of ML algorithms.

Even though each ML-based algorithm works on the same principle in its own way, it behaves differently for the data used. The ML algorithm must be trained and used for the specific data to be fit enough for classification or prediction. Hence, the models that were used are also considered for evaluation, as this can potentially provide a fair idea to the future researchers on the process of proceeding with their research in the domain of academic assessment and prediction.

Table 9. Summary of the features used.

Category	Feature	Category	Behavioral feature	Behavioral feature
Demographic Features	Preliminary education	Behavioral features	Personal	Number of times opened
	Gender		Self-evaluation	Number of times closed
	Age		Time management	Number of times "Next" used
	Location of stay		Anxiety	Number of times "Previous" used
	Duration of travel		Study aids	Number of times "seek" used
	Parent education		Study time duration	Number of times "jump" used
	Level of income		Isolation	Number of times "search" used
	Status of family		Search of emotional support	Activity and engagement
	Social support group		Self-blame	Number of forum replies
	Year of admission and age		Problem in focusing	Number of clarifications sought
	Number of siblings		Fatalism	Number of hand-raises
	Computer knowledge		Reaction time	Time spent online
	Type of parent employment		Avoiding amusement	Number of assignments submitted
	Type of student self-employment		Verbal communication	Number of tests submitted
	Disability		Interest and motivation	Time spent on assignment
	Mode of study		User navigation	Time spent on quiz
	Tuition fee source		Number of clicks on the discussion forum	Number of days absent
	Commuting		Number of clicks on material of study	Specificity of the days absent
Academic features	Individual semester grades		On-campus clicks versus off-campus clicks	Number of clicks on report
	Final exam grades		Number of clicks during weekdays	Number of clicks on mark issued
	Individual subject grades		Number of clicks during weekends	Dual pane activity
	Grade of previous semesters		Number of clicks on modules	E-books
	Oral exam grades		Number of bookmarks created	Number of times opened
	Written exam grades		Number of bookmarks deleted	Number of times closed
	Number of appearance for exams		Video content	
	Entrance test grades			
	Prerequisite course grade			
	Curriculum			
	Academic resource			

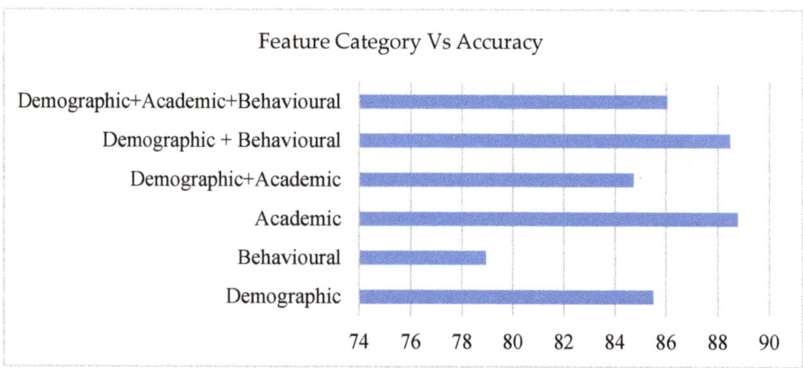

Figure 9. Accuracy against feature category.

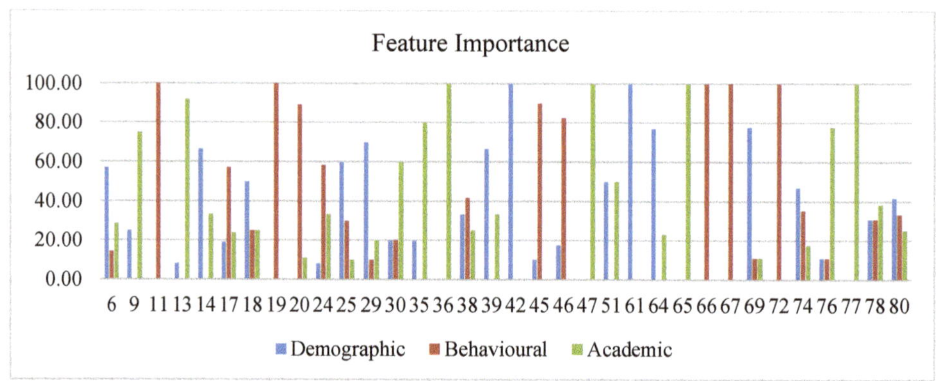

Figure 10. Feature importance in articles.

This systematic review was constrained to ML algorithms used in the domain of academic performance prediction. It was found that various algorithms have been used; some algorithms that were not used and applied were not considered for testing in this domain of research. Hence, the future researchers can consider those algorithms used in the previous studies as benchmarks and proceed with the unused ML models to showcase the results. Additionally, a bird's-eye view on the related disciplines of ML, namely AI, DL, statistics, and data mining with respect to this domain of research, can be done in order to devise new valuable ideas and provide the attempts to implement them.

4. Conclusions

Conclusions attained from the systematic review made are:

- DT and ensemble learning models have been employed in several selected articles, wherein NNs or transfer learning with appropriate layers can be adopted to make an unbiased decision on the model suitable for the collected data.
- Most articles focused only on a specific aspect of accuracy, and it seems to be a biased one. Indeed, the performance measures can be chosen from a wide variety of available measures suitable for the problem of study as classification or regression.
- The amount of data collected for the dataset can be computed in a high quantity and of a cohort nature of a specific set of students to analyze their change in behavioral features and demographic features that influences their academic feature study.
- Behavioral features were taken in a large quantity, which could be equated to the other two categories of features as academic and demographic features. In the online

mode of study, the demographic feature does not have much impact on the academic features, whereas during offline modes of study, three types of features contribute equally to the performance of the student, which, in turn, leads us to decide the dropout percentage.

When a model is proposed, it is a common practice to compare the performance of various ML models on the collected data, which can influence the correctness or credibility of the data collected. However, it is a perfect practice to compare the performance of the proposed model against the datasets that were used in already existing research studies to prove the precision of the model, which, in turn, may likely lead to fine tuning of the model to fit multiple datasets.

Author Contributions: Conceptualization, S.A. and P.B.; methodology, P.B.; software, A.Q.; validation, S.A. and A.Q.; formal analysis, M.M.; investigation, P.B.; resources, P.B. and M.M.; data curation, P.B.; writing—original draft preparation, P.B.; writing—review and editing, P.B. and S.A.; visualization, M.M.; supervision, S.A. and A.Q.; project administration, S.A. and A.Q.; funding acquisition, S.A., P.B., M.M., and A.Q. All authors have read and agreed to the published version of the manuscript.

Funding: The research and the APC are funded by King Khalid University under the research Grant No. RGP1/207/42.

Acknowledgments: We would like to acknowledge the King Khalid University for their immense support in every step of research work carried out.

Conflicts of Interest: The authors declare no conflict of interest.

Appendix A

Table A1. Estimation metrics (decision-tree-based models).

Ref.	Models Used	Accuracy	AUC	Recall	Precision	F1	Dataset Used	Quality Score
[11]	Random forest	77.29	NS	75.6	75.6	75.6	xAPI-Edu-Data	9
[12]	Random forest	86	68	86	85	85	Self-data from 3 different universities	9
[24]	Decision tree	79	NS	75.1	70.3	72	Webpage	9
[29]	Decision tree	98.94	99.4	100	85.7	92	Self-collected	9
[31]	Decision tree	95.82	NS	NS	NS	NS	UCI repository	8
[34]	Decision tree	96.5	NS	NS	93	NS	Collected data	9
[44]	Decision tree	75	NS	NS	NS	NS	Self-data	8
[46]	Decision tree	98.5	92.1	97.3	94	95.6	University of Stanford	9
[54]	Genetic algorithm-based decision tree	94.39	NS	NS	NS	NS	Federal Board of Pakistan	8
[55]	Random forest	79.8	93.8	79.8	78.8	79	University of Nigeria	7
[58]	Random forest	NS	93	NS	NS	NS	Self-data	9
[59]	Multiple linear regression	90	NS	90	89	89	Kaggle	7
[60]	Decision tree	87.21	NS	93.65	89.39	NS	University of Phayao	9
[62]	Random forest	70.1	NS	NS	NS	NS	University of Li'ege (Belgium)	9
[64]	Decision tree	94.63	NS	95.76	98.33	71.9	Open University of China	10
[79]	Decision tree	67.71	NS	NS	NS	NS	NEDUET, Pakistan	9

Note: NS—not specified.

Table A2. Estimation metrics (ensemble-based models).

Ref.	Models Used	Accuracy	Sensitivity	Specificity	AUC	Recall	Precision	F1	Dataset Used	Quality Score
[7]	Ensemble (J48, real AdaBoost)	95.78	NS	NS	NS	0.958	0.958	0.96	UCI Student Performance	8
[15]	Ensemble (reptree bagging)	97.5	NS	NS	NS	96.3	96.4	96.2	Self-data	8
[27]	Ensemble [DT, boosting]	96.96	NS	NS	NS	95.97	94.97	95.5	Self-data	8
[32]	Ensemble [NN, RF-boosting]	NS	NS	NS	NS	NS	NS	NS	NS	6
[43]	Ensemble [NB + AdaBoost]	98.12	96	100	NS	NS	NS	98	Directorate of Higher Secondary Education	7
[44]	Ensemble [DT-XGBoost]	NS	NS	NS	NS	92.5	89	89	OULAD	8
[47]	Ensemble [DT, SMO]	90.13	NS	NS	NS	NS	NS	NS	Microsoft showcase school "Avgoulea-Linardatou"	9
[57]	SVM-boosting	90.6	NS	NS	NS	NS	97	NS	North Carolina	5
[61]	[DT, ANN, SVM] Stacking ensemble	NS	NS	NS	77.7	74.52	NS	76.1	Self-data	10
[63]	Ensemble learning [random forest (RF) and adaptive boosting (AdaBoost)]	98	NS	NS	NS	91	69	78	Self-data	7
[66]	RF, KNN, and adaptive boosting	70	NS	NS	NS	70	70	79	University of León	9
[74]	Ensemble [RF, boosting]	98.22	NS	NS	NS	NS	NS	NS	Self-data	10
[76]	Hybrid linear vector quantization (LVQ + AdaBoost)	92.6	NS	NS	NS	95.6	91	92.3	NS	8
[78]	Ensemble (DT + K means clustering)	75.47	NS	NS	NS	72.2	47.27	57.1	NS	8
[80]	Ensemble learning (SVM, RF, AdaBoost + logistic regression via stacking)	NS	NS	NS	91.9	86	85.5	85	Hankou University	10

Note: NS—not specified.

Table A3. Estimation metrics (neural-network-based models).

Ref.	Models Used	Accuracy	Sensitivity	Specificity	AUC	Recall	Precision	F1	Dataset Used	Quality Score
[16]	NB, MLP, SMO, C4.5, JRip, kNN	85.43	NS	82.61	NS	97.48	NS	84.3	Self-data	10
[17]	MLP-BP(ANN)	100	NS	NS	100	100	100	100	Self-data	10
[22]	CNN	99.4	NS	NS	88.7	77.26	97	86	US K12 schools	8
[26]	Improved deep belief network	83.14			NS	NS	NS	NS	ADS, GT4M	10
[35]	NN	51.9	NS	NS	63.5	51.9	48.6	49.4	Self-data	10
[37]	NN	96	NS	NS	NS	92	96	89.2	NS	6
[39]	MLP(ANN)	94.8	NS	NS	NS	94.8	94.2	NS	STIKOM Poltek Cirebon	10
[45]	ANN	88.48	NS	NS	NS	69	93	NS	OULA	9
[56]	BPNN	84.8	94.8	54.6	NS	NS	86.3	NS	Self-data	6
[65]	MLR, MLP, RBF, SVM	89.9	NS	NS	NS	NS	NS	NS	NS	9
[67]	NN	96	NS	NS	NS	NS	NS	NS	University of Tartu in Estonia	10
[69]	NN—Levenberg–Marquardt learning algorithm	83.7	77.37	85.16	NS	NS	NS	NS	Self-data	10
[72]	RBF	76.92	100	60	NS	NS	NS	NS	NS	8
[77]	kNN	86	89	84	NS	NS	NS	81	Self-data	9

Note: NS—not specified.

Table A4. Estimation metrics (others-based models).

Ref.	Models Used	Accuracy	Sensitivity	Specificity	AUC	Recall	Precision	F1	Dataset Used	Quality Score
[3]	Adversarial network based deep support vector machine	0.954	0.971	0.968	NS	NS	NS	NS	Self-data	8
[10]	Multiple linear regression model	NS	NS	NS	NS	NS	NS	NS	Covenant University in Nigeria	6
[19]	LSTM	NS	NS	NS	68.2	NS	NS	NS	Canadian University	10
[20]	SVM	70.21	NS	NS	NS	NS	NS	NS	George Mason University	10
[25]	Decision tree, random Forest, support vector machine, logistic regression, AdaBoost, stochastic gradient descent	96.65	93.75	93.75	NS	NS	99.6	NS	UCI	10
[30]	Multiple regression algorithm	NS	NS	NS	NS	NS	NS	NS	Self-collected	10
[36]	Logistic regression	89.15	NS	NS	NS	NS	NS	NS	Covenant University	9
[40]	SVM	76.67	NS	NS	NS	NS	NS	NS		5
[41]	Non-linear SVM	NS	NS	NS	75	89	88	89	OULAD	7
[51]	LR	94.9	NS	NS	NS	NS	NS	NS	Imam Abdulrahman bin Faisal University	10
[53]	Vector-based SVM	93.8	94	93.6	NS	NS	NS	NS	OULA	7
[75]	Transfer learning (deep learning)	NS	NS	NS	NS	NS	NS	NS	NS	8

Note: NS—not specified.

References

1. Rebai, S.; Ben Yahia, F.; Essid, H. A graphically based machine learning approach to predict secondary schools performance in Tunisia. *Socio-Econ. Plan. Sci.* **2019**, *70*, 100724. [CrossRef]
2. Tatiana, A.C.; Cudney, E.A. Predicting Student Retention Using Support Vector Machines. *Procedia Manuf.* **2019**, *39*, 1827–1833. [CrossRef]
3. Chui, K.T.; Liu, R.W.; Zhao, M.; De Pablos, P.O. Predicting Students' Performance with School and Family Tutoring Using Generative Adversarial Network-Based Deep Support Vector Machine. *IEEE Access* **2020**, *8*, 86745–86752. [CrossRef]
4. Fernandez-Garcia, A.J.; Rodriguez-Echeverria, R.; Preciado, J.C.; Manzano, J.M.C.; Sanchez-Figueroa, F. Creating a Recommender System to Support Higher Education Students in the Subject Enrollment Decision. *IEEE Access* **2020**, *8*, 189069–189088. [CrossRef]
5. Xu, J.; Moon, K.H.; van der Schaar, M. A Machine Learning Approach for Tracking and Predicting Student Performance in Degree Programs. *IEEE J. Sel. Top. Signal Process.* **2017**, *11*, 742–753. [CrossRef]
6. Song, X.; Li, J.; Sun, S.; Yin, H.; Dawson, P.; Doss, R.R.M. SEPN: A Sequential Engagement Based Academic Performance Prediction Model. *IEEE Intell. Syst.* **2020**, *36*, 46–53. [CrossRef]
7. Imran, M.; Latif, S.; Mehmood, D.; Shah, M.S. Student Academic Performance Prediction using Supervised Learning Techniques. *Int. J. Emerg. Technol. Learn. (iJET)* **2019**, *14*, 92–104. [CrossRef]
8. Rivera, J.E.H. A Hybrid Recommender System to Enrollment for Elective Subjects in Engineering Students using Classification Algorithms. *Int. J. Adv. Comput. Sci. Appl.* **2020**, *11*, 400–406. [CrossRef]
9. Cen, L.; Ruta, D.; Powell, L.; Hirsch, B.; Ng, J. Quantitative approach to collaborative learning: Performance prediction, individual assessment, and group composition. *Int. J. Comput. -Supported Collab. Learn.* **2016**, *11*, 187–225. [CrossRef]
10. Adekitan, A.I.; Shobayo, O. Gender-based comparison of students' academic performance using regression models. *Eng. Appl. Sci. Res.* **2020**, *47*, 241–248.
11. Enaro, A.O.; Chakraborty, S. Feature Selection Algorithms for Predicting Students Academic Performance Using Data Mining Techniques. *Int. J. Sci. Technol. Res.* **2020**, *9*, 3622–3626.
12. Huang, A.Y.Q.; Lu, O.H.T.; Huang, J.C.H.; Yin, C.J.; Yang, S.J.H. Predicting Students' Academic Performance by Using Educational Big Data and Learning Analytics: Evaluation of Classification Methods and Learning Logs. *Interact. Learn. Environ.* **2020**, *28*, 206–230. [CrossRef]
13. Xu, X.; Wang, J.; Peng, H.; Wu, R. Prediction of academic performance associated with internet usage behaviors using machine learning algorithms. *Comput. Hum. Behav.* **2019**, *98*, 166–173. [CrossRef]
14. Livieris, I.E.; Drakopoulou, K.; Tampakas, V.T.; Mikropoulos, T.A.; Pintelas, P. Predicting Secondary School Students' Performance Utilizing a Semi-supervised Learning Approach. *J. Educ. Comput. Res.* **2018**, *57*, 448–470. [CrossRef]

15. Shanthini, A.; Vinodhini, G.; Chandrasekaran, R.M. Predicting Students' Academic Performance in the University Using Meta Decision Tree Classifiers. *J. Comput. Sci* **2018**, *14*, 654–662. [CrossRef]

16. Vialardi, C.; Chue, J.; Peche, J.; Alvarado, G.; Vinatea, B.; Estrella, J.; Ortigosa, Á. A data mining approach to guide students through the enrollment process based on academic performance. *User Model User-Adap. Inter.* **2011**, *21*, 217–248. [CrossRef]

17. Musso, M.F.; Hernández, C.F.R.; Cascallar, E.C. Predicting key educational outcomes in academic trajectories: A machine-learning approach. *High. Educ.* **2020**, *80*, 875–894. [CrossRef]

18. Lagman, A.C.; Alfonso, L.P.; Goh, M.L.I.; Lalata, J.-A.P.; Magcuyao, J.P.H.; Vicente, H.N. Classification Algorithm Accuracy Improvement for Student Graduation Prediction Using Ensemble Model. *Int. J. Inf. Educ. Technol.* **2020**, *10*, 723–727. [CrossRef]

19. Chen, F.; Cui, Y. Utilizing Student Time Series Behaviour in Learning Management Systems for Early Prediction of Course Performance. *J. Learn. Anal.* **2020**, *7*, 1–17. [CrossRef]

20. Damuluri, S.; Islam, K.; Ahmadi, P.; Qureshi, N.S. Analyzing Navigational Data and Predicting Student Grades Using Support Vector Machine. *Emerg. Sci. J.* **2020**, *4*, 243–252. [CrossRef]

21. Kabakus, A.T.; Senturk, A. An analysis of the professional preferences and choices of computer engineering students. *Comput. Appl. Eng. Educ.* **2020**, *28*, 994–1006. [CrossRef]

22. Yang, Z.; Yang, J.; Rice, K.; Hung, J.-L.; Du, X. Using Convolutional Neural Network to Recognize Learning Images for Early Warning of At-Risk Students. *IEEE Trans. Learn. Technol.* **2020**, *13*, 617–630. [CrossRef]

23. Cruz-Jesus, F.; Castelli, M.; Oliveira, T.; Mendes, R.; Nunes, C.; Sa-Velho, M.; Rosa-Louro, A. Using artificial intelligence methods to assess academic achievement in public high schools of a European Union country. *Heliyon* **2020**, *6*, e04081. [CrossRef] [PubMed]

24. Figueroa-Canas, J.; Sancho-Vinuesa, T. Early Prediction of Dropout and Final Exam Performance in an Online Statistics Course. *IEEE Rev. Iberoam. De Tecnol. Del Aprendiz.* **2020**, *15*, 86–94. [CrossRef]

25. Razaque, A.; Alajlan, A. Supervised Machine Learning Model-Based Approach for Performance Prediction of Students. *J. Comput. Sci.* **2020**, *16*, 1150–1162. [CrossRef]

26. Sokkhey, P.; Okazaki, T. Study on Dominant Factor for Academic Performance Prediction using Feature Selection Methods. *Int. J. Adv. Comput. Sci. Appl.* **2020**, *11*. [CrossRef]

27. Almasri, A.; Alkhawaldeh, R.S.; Çelebi, E. Clustering-Based EMT Model for Predicting Student Performance. *Arab. J. Sci. Eng.* **2020**, *45*, 10067–10078. [CrossRef]

28. Sethi, K.; Jaiswal, V.; Ansari, M. Machine Learning Based Support System for Students to Select Stream (Subject). *Recent Adv. Comput. Sci. Commun.* **2020**, *13*, 336–344. [CrossRef]

29. Gil, J.S.; Delima, A.J.P.; Vilchez, R.N. Predicting Students' Dropout Indicators in Public School using Data Mining Approaches. *Int. J. Adv. Trends Comput. Sci. Eng.* **2020**, *9*, 774–778. [CrossRef]

30. Qazdar, A.; Er-Raha, B.; Cherkaoui, C.; Mammass, D. A machine learning algorithm framework for predicting students performance: A case study of baccalaureate students in Morocco. *Educ. Inf. Technol.* **2019**, *24*, 3577–3589. [CrossRef]

31. Gamao, A.O.; Gerardo, B.D.; Medina, R.P. Prediction-Based Model for Student Dropouts using Modified Mutated Firefly Algorithm. *Int. J. Adv. Trends Comput. Sci. Eng.* **2019**, *8*, 3461–3469. [CrossRef]

32. Susheelamma, H.K.; Ravikumar, K.M. Student risk identification learning model using machine learning approach. *Int. J. Electr. Comput. Eng.* **2019**, *9*, 3872–3879. [CrossRef]

33. Kostopoulos, G.; Kotsiantis, S.; Fazakis, N.; Koutsonikos, G.; Pierrakeas, C. A Semi-Supervised Regression Algorithm for Grade Prediction of Students in Distance Learning Courses. *Int. J. Artif. Intell. Tools* **2019**, *28*. [CrossRef]

34. Buenaño-Fernández, D.; Gil, D.; Luján-Mora, S. Application of Machine Learning in Predicting Performance for Computer Engineering Students: A Case Study. *Sustainability* **2019**, *11*, 2833. [CrossRef]

35. Adekitan, A.I.; Noma-Osaghae, E. Data mining approach to predicting the performance of first year student in a university using the admission requirements. *Educ. Inf. Technol.* **2019**, *24*, 1527–1543. [CrossRef]

36. Adekitan, A.I.; Salau, O. The impact of engineering students' performance in the first three years on their graduation result using educational data mining. *Heliyon* **2019**, *5*, e01250. [CrossRef]

37. Maitra, S.; Eshrak, S.; Bari, M.A.; Al-Sakin, A.; Munia, R.H.; Akter, N.; Haque, Z. Prediction of Academic Performance Applying NNs: A Focus on Statistical Feature-Shedding and Lifestyle. *Int. J. Adv. Comput. Sci. Appl.* **2019**, *10*, 561–570. [CrossRef]

38. Almasri, A.; Celebi, E.; Alkhawaldeh, R. EMT: Ensemble Meta-Based Tree Model for Predicting Student Performance. *Sci. Program.* **2019**, *2019*, 3610248. [CrossRef]

39. Nurhayati, O.D.; Bachri, O.S.; Supriyanto, A.; Hasbullah, M. Graduation Prediction System Using Artificial Neural Network. *Int. J. Mech. Eng. Technol.* **2018**, *9*, 1051–1057.

40. Aluko, R.O.; Daniel, E.I.; Oshodi, O.S.; Aigbavboa, C.O.; Abisuga, A.O. Towards reliable prediction of academic performance of architecture students using data mining techniques. *J. Eng. Des. Technol.* **2018**, *16*, 385–397. [CrossRef]

41. Nadar, N.; Kamatchi, R. A Novel Student Risk Identification Model using Machine Learning Approach. *Int. J. Adv. Comput. Sci. Appl.* **2018**, *9*, 305–309. [CrossRef]

42. Kostopoulos, G.; Kotsiantis, S.; Pierrakeas, C.; Koutsonikos, G.; Gravvanis, G.A. Forecasting students' success in an open university. *Int. J. Learn. Technol.* **2018**, *13*, 26–43. [CrossRef]

43. Navamani, J.M.A.; Kannammal, A. Predicting performance of schools by applying data mining techniques on public examination results. *Res. J. Appl. Sci. Eng. Technol.* **2015**, *9*, 262–271. [CrossRef]

44. Wakelam, E.; Jefferies, A.; Davey, N.; Sun, Y. The potential for student performance prediction in small cohorts with minimal available attributes. *Br. J. Educ. Technol.* **2019**, *51*, 347–370. [CrossRef]
45. Waheed, H.; Hassan, S.U.; Aljohani, N.R.; Hardman, J.; Alelyani, S.; Nawaz, R. Predicting academic performance of students from VLE big data using deep learning models. *Comput. Hum. Behav.* **2020**, *104*, 106189. [CrossRef]
46. Mourdi, Y.; Sadgal, M.; El Kabtane, H.; Fathi, W.B.; El Kabtane, H.; Youssef, M.; Mohamed, S.; Wafaa, B.F. A machine learning-based methodology to predict learners' dropout, success or failure in MOOCs. *Int. J. Web Inf. Syst.* **2019**, *15*, 489–509. [CrossRef]
47. Livieris, I.E.; Kotsilieris, T.; Tampakas, V.; Pintelas, P. Improving the evaluation process of students' performance utilizing a decision support software. *Neural Comput. Appl.* **2018**, *31*, 1683–1694. [CrossRef]
48. Son, L.H.; Fujita, H. Neural-fuzzy with representative sets for prediction of student performance. *Appl. Intell.* **2019**, *49*, 172–187. [CrossRef]
49. Coussement, K.; Phan, M.; De Caigny, A.; Benoit, D.F.; Raes, A. Predicting student dropout in subscription-based online learning environments: The beneficial impact of the logit leaf model. *Decis. Support Syst.* **2020**, *135*, 113325. [CrossRef]
50. Injadat, M.; Moubayed, A.; Nassif, A.B.; Shami, A. Systematic ensemble model selection approach for educational data mining. *Knowl.-Based Syst.* **2020**, *200*, 105992. [CrossRef]
51. Tatar, A.E.; Düştegör, D. Prediction of Academic Performance at Undergraduate Graduation: Course Grades or Grade Point Average? *Appl. Sci.* **2020**, *10*, 4967. [CrossRef]
52. Karthikeyan, V.G.; Thangaraj, P.; Karthik, S. Towards developing hybrid educational data mining model (HEDM) for efficient and accurate student performance evaluation. *Soft Comput.* **2020**, *24*, 18477–18487. [CrossRef]
53. Chui, K.T.; Fung, D.C.L.; Lytras, M.D.; Lam, T.M. Predicting at-risk university students in a virtual learning environment via a machine learning algorithm. *Comput. Hum. Behav.* **2020**, *107*, 105584. [CrossRef]
54. Yousafzai, B.K.; Hayat, M.; Afzal, S. Application of machine learning and data mining in predicting the performance of intermediate and secondary education level student. *Educ. Inf. Technol.* **2020**, *25*, 4677–4697. [CrossRef]
55. Adekitan, A.I.; Salau, O. Toward an improved learning process: The relevance of ethnicity to data mining prediction of students' performance. *SN Appl. Sci.* **2020**, *2*, 8. [CrossRef]
56. Lau, E.T.; Sun, L.; Yang, Q. Modelling, prediction and classification of student academic performance using artificial neural networks. *SN Appl. Sci.* **2019**, *1*, 982. [CrossRef]
57. Sorensen, L.C. "Big Data" in Educational Administration: An Application for Predicting School Dropout Risk. *Educ. Adm. Q.* **2019**, *55*, 404–446. [CrossRef]
58. Alsuwaiket, M.; Blasi, A.H.; Al-Msie'deen, R.F. Formulating module assessment for improved academic performance predictability in higher education. *arXiv* **2020**, arXiv:2008.13255.
59. Suguna, R.; Shyamala Devi, M.; Bagate, R.A.; Joshi, A.S. Assessment of feature selection for student academic performance through machine learning classification. *J. Stat. Manag. Syst.* **2019**, *22*, 729–739. [CrossRef]
60. Nuankaew, P. Dropout Situation of Business Computer Students, University of Phayao. *Int. J. Emerg. Technol. Learn.* **2019**, *14*, 115–131. [CrossRef]
61. Adejo, O.W.; Connolly, T. Predicting student academic performance using multi-model heterogeneous ensemble approach. *J. Appl. Res. High. Educ.* **2018**, *10*, 61–75. [CrossRef]
62. Hoffait, A.-S.; Schyns, M. Early detection of university students with potential difficulties. *Decis. Support Syst.* **2017**, *101*, 1–11. [CrossRef]
63. Rovira, S.; Puertas, E.; Igual, L. Data-driven system to predict academic grades and dropout. *PLoS ONE* **2017**, *12*, 0171207. [CrossRef] [PubMed]
64. Tan, M.; Shao, P. Prediction of student dropout in e-Learning program through the use of machine learning method. *Int. J. Emerg. Technol. Learn.* **2015**, *10*, 11. [CrossRef]
65. Huang, S.; Fang, N. Predicting student academic performance in an engineering dynamics course: A comparison of four types of predictive mathematical models. *Comput. Educ.* **2013**, *61*, 133–145. [CrossRef]
66. Guerrero-Higueras, Á.M.; Fernández Llamas, C.; Sánchez González, L.; Gutierrez Fernández, A.; Esteban Costales, G.; González, M.Á.C. Academic Success Assessment through Version Control Systems. *Appl. Sci.* **2020**, *10*, 1492. [CrossRef]
67. Hooshyar, D.; Pedaste, M.; Yang, Y. Mining Educational Data to Predict Students' Performance through Procrastination Behavior. *Entropy* **2020**, *22*, 12. [CrossRef]
68. Ezz, M.; Elshenawy, A. Adaptive recommendation system using machine learning algorithms for predicting student's best academic program. *Educ. Inf. Technol.* **2019**, *25*, 2733–2746. [CrossRef]
69. Al-Sudani, S.; Palaniappan, R. Predicting students' final degree classification using an extended profile. *Educ. Inf. Technol.* **2019**, *24*, 2357–2369. [CrossRef]
70. Gray, C.C.; Perkins, D. Utilizing early engagement and machine learning to predict student outcomes. *Comput. Educ.* **2019**, *131*, 22–32. [CrossRef]
71. Garcia, J.D.; Skrita, A. Predicting Academic Performance Based on Students' Family Environment: Evidence for Colombia Using Classification Trees. *Psychol. Soc. Educ.* **2019**, *11*, 299–311. [CrossRef]
72. Babić, I.Đ. Machine learning methods in predicting the student academic motivation. *Croat. Oper. Res. Rev.* **2017**, *8*, 443–461. [CrossRef]

73. Kotsiantis, S.B. Use of machine learning techniques for educational proposes: A decision support system for forecasting students' grades. *Artif. Intell. Rev.* **2012**, *37*, 331–344. [CrossRef]
74. Sokkhey, P.; Okazaki, T. Hybrid Machine Learning Algorithms for Predicting Academic Performance. *Int. J. Adv. Comput. Sci. Appl.* **2020**, *11*, 32–41. [CrossRef]
75. Hussain, S.; Gaftandzhieva, S.; Maniruzzaman, M.; Doneva, R.; Muhsin, Z.F. Regression analysis of student academic performance using deep learning. *Educ. Inf. Technol.* **2020**, *26*, 783–798. [CrossRef]
76. Bhagavan, K.S.; Thangakumar, J.; Subramanian, D.V. Predictive analysis of student academic performance and employability chances using HLVQ algorithm. *J. Ambient. Intell. Humaniz. Comput.* **2020**, *12*, 3789–3797. [CrossRef]
77. Akçapınar, G.; Altun, A.; Aşkar, P. Using learning analytics to develop early-warning system for at-risk students. *Int. J. Educ. Technol. High. Educ.* **2019**, *16*, 40. [CrossRef]
78. Francis, B.K.; Babu, S.S. Predicting Academic Performance of Students Using a Hybrid Data Mining Approach. *J. Med. Syst.* **2019**, *43*, 162. [CrossRef]
79. Asif, R.; Hina, S.; Haque, S.I. Predicting student academic performance using data mining methods. *Int. J. Comput. Sci. Netw. Secur.* **2017**, *17*, 187–191.
80. Yan, L.; Liu, Y. An Ensemble Prediction Model for Potential Student Recommendation Using Machine Learning. *Symmetry* **2020**, *12*, 728. [CrossRef]
81. Nicolas, P.R. *Leverage Scala and Machine Learning to Construct and Study Systems that Can Learn from Data*; Packt Publishing Ltd.: Birmingham, UK, 2015.

*applied sciences*

Article

A Novel Method for Performance Measurement of Public Educational Institutions Using Machine Learning Models

Talha Mahboob Alam [1,*,†], Mubbashar Mushtaq [2,†], Kamran Shaukat [3,4,*], Ibrahim A. Hameed [5,*], Muhammad Umer Sarwar [6] and Suhuai Luo [3]

[1] Department of Computer Science and Information Technology, Virtual University of Pakistan, Lahore 54890, Pakistan
[2] Department of Computer Science, University of Engineering and Technology, Lahore 54890, Pakistan; mubashar287@gmail.com
[3] School of Information and Physical Sciences, The University of Newcastle, Callaghan 2308, Australia; suhuai.luo@newcastle.edu.au
[4] Department of Data Science, University of the Punjab, Lahore 54890, Pakistan
[5] Department of ICT and Natural Sciences, Norwegian University of Science and Technology, 7491 Trondheim, Norway
[6] Department of Computer Science, Government College University Faisalabad, Faisalabad 38000, Pakistan; sarwaromer@gmail.com
* Correspondence: Talhamahboob95@gmail.com (T.M.A.); Kamran.shaukat@uon.edu.au (K.S.); ibib@ntnu.no (I.A.H.)
† Talha Mahboob Alam and Mubbashar Mushtaq contributed equally to this work.

Citation: Alam, T.M.; Mushtaq, M.; Shaukat, K.; Hameed, I.A.; Umer Sarwar, M.; Luo, S. A Novel Method for Performance Measurement of Public Educational Institutions Using Machine Learning Models. *Appl. Sci.* **2021**, *11*, 9296. https://doi.org/10.3390/app11199296

Academic Editor: Carlos A. Iglesias

Received: 14 July 2021
Accepted: 30 September 2021
Published: 7 October 2021

Publisher's Note: MDPI stays neutral with regard to jurisdictional claims in published maps and institutional affiliations.

Abstract: Lack of education is a major concern in underdeveloped countries because it leads to poor human and economic development. The level of education in public institutions varies across all regions around the globe. Current disparities in access to education worldwide are mostly due to systemic regional differences and the distribution of resources. Previous research focused on evaluating students' academic performance, but less has been done to measure the performance of educational institutions. Key performance indicators for the evaluation of institutional performance differ from student performance indicators. There is a dire need to evaluate educational institutions' performance based on their disparities and academic results on a large scale. This study proposes a model to measure institutional performance based on key performance indicators through data mining techniques. Various feature selection methods were used to extract the key performance indicators. Several machine learning models, namely, J48 decision tree, support vector machines, random forest, rotation forest, and artificial neural networks were employed to build an efficient model. The results of the study were based on different factors, i.e., the number of schools in a specific region, teachers, school locations, enrolment, and availability of necessary facilities that contribute to school performance. It was also observed that urban regions performed well compared to rural regions due to the improved availability of educational facilities and resources. The results showed that artificial neural networks outperformed other models and achieved an accuracy of 82.9% when the relief-F based feature selection method was used. This study will help support efforts in governance for performance monitoring, policy formulation, target-setting, evaluation, and reform to address the issues and challenges in education worldwide.

Keywords: performance measurement; key performance indicators; educational data mining; institutes performance; governance

1. Introduction

The education system enhances nation-building, reduces poverty, and promotes learning opportunities [1]. Children's education is essential for economic development. Past research revealed that initial schooling and the living environment significantly impact an individual's personality and education [2]. Human capital is a fundamental resource for a country's economic growth. Massive public investment in education facilitates human capital formation, which returns rewards in the form of higher productivity, higher wages, and financial growth [3,4]. Students' academic performance plays a vital role in the generation of a qualified professional workforce, which is responsible for the country's social and economic development. Student academic performance has attracted substantial attention in past research [5]. Student performance is based on personal, social, economic, psychological, and environmental factors. Most researchers used student results or grade point averages (GPA) to evaluate the individual performance. Various studies also considered teachers' education, family background, gender, class environment, class size, lesson plans, reading materials, innovation in class, examination frameworks, family, work, and extracurricular activities [6]. The distribution of resources strongly affects the performance of rural and urban students, and mostly rural students appear to be deprived. This implies differences in student academic performance and other social outcomes such as intelligence, aspirations, grooming, motivation, and aptitude. Rural–urban inequality in academic performance remains challenging and unresolved and has become a global issue [2]. School performance varies among different regions and groups due to differences in educational opportunities. These variations in opportunities and achievement have become a global concern, especially for developing countries [1,4], and such problems have also been recorded in emerging regions for female students [1]. The quality of education has been declining across Pakistan, including Punjab. Conditions in public schools are not satisfactory, especially since the academic outcomes of students in rural areas are poor compared to those in the country's urban regions [7]. The insufficient allocation of resources for education and a large budget deficit, especially in developing countries such as Pakistan, decrease school performance and present a challenge for policymakers. Limited studies on educational inputs and output in Pakistan mostly focused on specific regions [3,7]. One study conducted in Khyber Pakhtunkhwa revealed that essentials of educational infrastructure such as teaching quality, drinking water, gas, electricity, and school building conditions positively impact educational outcomes [3].

Researchers have recognised the impact of the surrounding environment on the performance of academic institutions [1,4,8]. Mostly, their focus remained on student academic performance [8,9], but some researchers targeted a particular region on a small scale [3,7]. Some research focused on early predictors of student success rates in higher education institutions (HEIs) [10]. Some studies considered basic facility parameters (i.e., electricity, gas, libraries, and teaching quality) and showed their impact on schools in some districts of Khyber Pakhtunkhwa province, in Pakistan [3]. Educational opportunities reflect the local school environment and socio-economic factors [1] because the performance of educational institutions in urban areas is different from that in rural areas [1,2,4]. The disparities in school education are due to regional differences and gender inequalities across Turkey [11]. The association of different parameters in different country regions was analysed. In some regions, the number of females in the local population is higher because males tend to migrate earlier for employment.

Moreover, institutional facilities and learning environments directly affect the performance of school institutions. Punjab's school education department conducts quarterly district rankings to track school performance and timely highlight those schools that are lagging. This ranking is based on various indicators such as student attendance, teacher presence, and the availability of boundary walls, toilets, drinking water, and furniture. The ranking statistics still show the need to uplift educational levels in different districts of Punjab [12]. Discovering new information from a massive amount of data is challenging and sometimes too expensive [8]. The most commonly adopted process used to extract hidden information from a large amount of data is data mining (DM). The approach used to extract meaningful knowledge from educational data is known as education data mining (EDM). Different machine learning-based models are used for performance measurement, including random forest, decision tree, K-nearest neighbour, and naïve Bayes [13]. This study proposes a framework to measure the institutional performance based on key performance indicators through data mining techniques.

Contributions: This study offers several contributions in the education domain to measure the performance of educational institutions.

1. A state-of-the-art dataset has been collected regarding the different indicators to measure the performance of educational institutions. The collected dataset was prone to noise, biases, and missing and outlier values.
2. Much work has been done to evaluate individual schools or measure student performance rather than institutional performance. To the best of our knowledge, no work has been done to measure institutional performance. However, a novel method for performance measurement of public institutions through machine learning models has been proposed in this study.
3. Regarding institutional performance, a regional perspective has been applied. This indicator has not been explored in the literature to investigate the performance of institutions.
4. Significant feature selection techniques were combined with machine learning models to develop the proposed framework for the performance measurement of public schools. It has also been observed that differences in demographics and provided facilities emerged due to regional differences.
5. This study will help support governance for performance monitoring, policy formulation, target-setting, evaluation, and reforms. The achieved results will help to address the issues and challenges in education worldwide.

The rest of the paper is organised as follows: Section 2 provides the related work. Section 3 describes the proposed methodology of our implementation methods. Section 4 presents the implementation results, while Section 5 analyses the results and implications of our study. Finally, we state our conclusions in Section 6.

2. Literature Review

Jamil et al. [3] explained the effect of institutional factors on student educational performance. The research was carried out on a large dataset consisting of 1642 schools in Khyber Pakhtunkhwa province, Pakistan. A positive relationship was found between student performance and institutional factors such as the availability of electricity, gas, and library facilities. In rural areas, electricity and gas had a positive impact, and well-constructed schools improved students' outcomes in these areas. However, factors such as infrastructure and teaching quality were not considered in their study. Tesema and Braeken [1] investigated students' educational achievement in terms of regional and gender differences. The regional differences were based on socio-economic and school environment-related factors. The analysis examined 2 years of grade 12 results. The results in developed regions were found to be better compared to those in emerging regions.

The results also revealed that those regions where the gender gap was minimal had a higher education rate than those with a high gender gap. But their study only considered one district, which may not be generalised. Eduardo Fernandes et al. [8] presented a predictive analysis of students in public schools in terms of their academic performance. A data mining classification model, gradient boosting machine (GBM), was implemented to predict student academic outcomes at the end of the year. The results showed the most significant attributes for prediction were students' grades and their class absence rates. Moreover, other important attributes such as the school medium, school segregation by gender, and the number of teachers were also crucial.

Gumus and Chudgar [11] concluded that unschooled children were a consequence of regional differences and gender inequalities. The analytical approach of binary logistic regression was applied to the dataset. The results indicated that student demographic characteristics such as gender, age, and home factors such as parent education and family financial status were significantly associated with students' school participation. Their study was limited in the perspective of the impact of regional dimensions on student performance. It concluded that disparities among regions must be considered in terms of socio-economic, demographic, and geographic factors that affect school participation. Nurliana and Sudaryana [14] investigated the factors that improve the student learning process and increase student knowledge. The experiment was performed on the dataset of students and teachers at one school for 1 year. Some students were taught using the old, traditional methods while other students were taught with the latest methods and proper equipment and facilities. The behaviour and interest of students revealed that better instructional tools and facilities increase the interest of students toward learning. However, they could not be considered key factors like number of students, number of classrooms, or availability of classrooms.

Hameen et al. [15] considered school facilities factors and determined their impact on student attendance, academic performance, and health. Their research covered schools in the United States. The analysis showed that schools with good classroom heating facilities and air conditioning for the summer season had a high attendance rate compared to schools that lacked these facilities. It was concluded that investments in school mechanical and plumbing systems improve student health and lead to better academic outcomes. Their study did not consider the availability of playgrounds in the schools. Belmonte et al. [16] explored the impact of investments in school infrastructure on student outcomes. The research was conducted with data on high schools that received extra funds following the 2012 earthquake in Italy. Their approach utilized a quasi-experimental design and an instrumental variable strategy. Variations in the distribution of funds were noted. The results revealed that spending more on school infrastructure improves student outcomes. A better learning environment boosts motivation to study, in turn increasing student achievement. Gul and Farooq [17] highlighted the World Health Organization (WHO) guidelines for developing countries such as Pakistan to improve access to the physical environment of schools. The analysis was performed on schools in one region, Multan. A questionnaire approach was adopted for analysis purposes and to obtain feedback from school teaching staff. The questionnaire consisted of 10 core indicators. The 10 indicators were water facilities, water quantity, water quality, hygiene promotion practices, control of vector-borne diseases, toilet and handwashing facilities, cleaning and waste disposal systems, school safety, school building conditions, and supportive classroom conditions. Based on the analysis results, it was concluded that schools did not meet these 10 core indicators due to a district score (1.01) that was below the WHO's recommended score (1.5). These deficiencies were causes of poor student performance outcomes and had a negative impact on student health. However, the researchers only considered one district in their study. An overview of existing research techniques is presented in Table 1.

Table 1. Overview of existing techniques.

Reference	Year	Dataset	Machine Learning Technique	Feature Selection Technique	Institutional Performance Evaluation
[3]	2018	1642 Schools, Pakistan	✗	✗	✓
[1]	2018	NAEA 2014 Data	✗	✗	✓
[8]	2019	Brazil One Region School	✓	✗	✗
[11]	2016	TDHS-2008 Survey, Turkey	✗	✗	✓
[14]	2020	Vocational High School, Indonesia.	✗	✗	✗
[15]	2020	Data of US 125 Schools	✗	✗	✗
[16]	2019	INVALSI	✗	✗	✓
[17]	2019	158 Schools of District Multan, Pakistan	✗	✗	✗
Our work	2020	6674 high schools of Punjab	✓	✓	✓

3. Methodology

The traditional cross-industry standard process for data mining (CRISP-DM) was utilised to predict the performance of schools, as shown in Figure 1. The methodology consists of dataset collection, data preparation, modelling, and validation of results. Firstly, Punjab annual census data were obtained from the official website. Secondly, data preparation techniques were applied, i.e., data cleaning, data transformation, data normalisation, and discretisation. Thirdly, various feature selection techniques were utilised to extract significant features. Fourthly, various machine learning classifiers were employed to train the model. Lastly, different performance measures were utilised to check the performance of classifiers. Microsoft Azure Machine Learning Studio and WEKA were utilised for data analysis, preparation, and modelling.

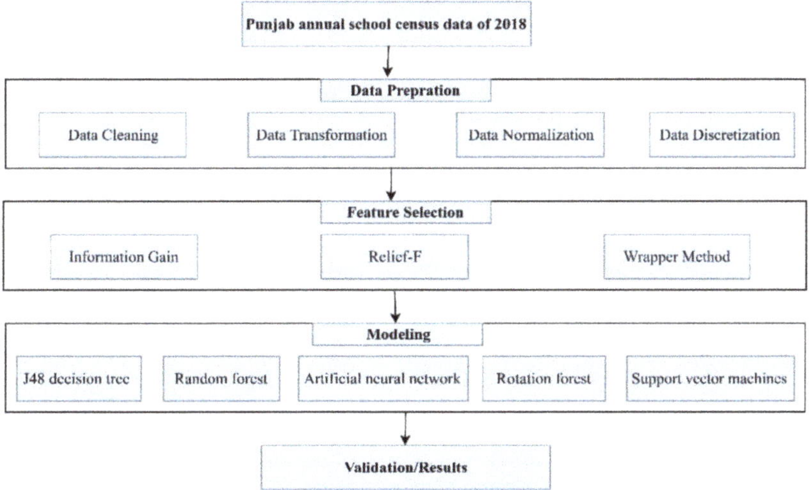

Figure 1. Proposed methodology.

3.1. Dataset

In Pakistan, students are awarded a secondary school certificate (SSC) after completing 10 years of schooling, also known as Matric. So, the study was delimited to 10th grade students in public schools. The dataset contained 108 attributes related to student enrolment, employee availability, location, the status of school basic facilities, and student outcomes (Matric results). The raw file is available at http://dx.doi.org/10.17632/637d4s7vjh.1. A few attributes in the raw data, e.g., school gender, library presence, school shift, school medium, and building condition were categorical attributes that contained various categories. Furthermore, many attributes, e.g., total schools, total urban schools, total computer labs, and total available classrooms, students without furniture, open-air class sections, students with furniture, total rural schools, total students, total playgrounds, total science labs, school with electricity facility, deficiency of classrooms, and total teachers were continuous attributes. These attributes contained numerical values after data pre-processing (data discretisation); continuous attributes have were converted into categorical (specified range) attributes. The target attribute was categorised into three classes based on institution-wise Matric marks: below 50% for low, between 51 and 70% for medium, and more than 70% for high. The *Matric result* or class attribute contained three values, i.e., high, medium, and low.

3.2. Data Preparation

This covers the steps related to the preparation of the dataset from raw data. Data preparation tasks are often performed repeatedly and not in any predefined order. These tasks include data cleaning, data normalisation, outlier detection, data reduction, and data transformation. Data preparation aids in generating a good model that may help to obtain effective results.

3.2.1. Data Cleaning

In data cleaning, redundant instances are detected and removed from the data. Data cleaning includes outlier detection and missing values imputation. Certain attributes observed to contain missing values, such as open-air class sections, total functional classrooms, are replaced with median values by using Equations (1) and (2):

$$For\ odd\ data\ elements = \frac{(n+1)}{2}\ th\ term \tag{1}$$

$$For\ even\ data\ elements = \frac{\frac{n}{2}\ th\ term + \left(\frac{n}{2}+1\right)\ th\ term}{2} \tag{2}$$

Outliers are those extreme values that show extreme deviation from mean values of the data, which can cause an error. Some negative values were observed in the "students without furniture" attribute, and was replaced with 0 after comparing and analysing the other instances.

3.2.2. Data Transformation

Data transformation has a meaningful effect on data mining since it helps fix the missing values in the data and brings information to the surface by creating new features to represent trends and other ratios. Some features, such as total playground, schools with electricity facilities, total computer labs, and total science labs, held values in Yes and No, which were converted to 0, 1. The attribute "deficiency of classrooms" was calculated based on available classrooms and by considering the general formula of one room for 40 students as stated in Equation (3):

$$Deficiency\ of\ Classrooms = \frac{Total\ Enrollment}{40} - Available\ Functional\ Classrooms \tag{3}$$

The attribute "Students without furniture" was calculated based on the "Students with furniture" and "Total Enrolment" as described in Equation (4):

$$Student\ without\ furniture = Total\ Enrollment - Student\ with\ furniture \tag{4}$$

The attribute "school location" was further split into two attributes (rural, urban) based on type of area. The data was converted into tehsil wise by aggregating values (by applying sum, count, average functions) to prepare the attributes such as total school, total teachers, total students, total rural schools, total urban schools, open-air class sections, total computer labs, total science labs, total playgrounds, total available classrooms, deficiency of classrooms, students with furniture, students without furniture, schools with electricity facilities, and Matric result.

3.2.3. Data Normalization

When multiple attributes have different scales, results may be affected. Normalisation brings all attributes to the same scale. All attributes were scaled into smaller ranges between 0 and 1. All integer attributes such as total computer labs, total science labs, and playgrounds were scaled between 0–1. The most common normalisation method is the Min-Max normalisation, used in this study. Furthermore, the Min-Max technique is efficient because results may be enhanced when data have outliers or missing values, as in our dataset [18]. This technique scaled all the numerical values of a numerical feature to a specified range and computed them through Equation (5).

$$X_{norm} = \frac{X - X_{min}}{X_{max} - X_{min}} \tag{5}$$

3.2.4. Data Discretisation

In data discretisation, numeric data are transformed by mapping values to interval or concept labels. This could be achieved using various techniques such as binning, correlation, cluster, and decision tree analysis. The binning method was utilised for data discretisation in this study. Additionally, the equal-frequency interval-based discretisation method was employed. In this method, the minimum and maximum values of all discretised attributes are determined. Then, these values are sorted in ascending order. The sorted values are further divided into k intervals, as each interval contains n/k data instances. There may occur continuous value, which can cause the occurrence to be assigned into different bins. The limitation of equal width interval discretisation is overcome by adopting the domain's approach according to the same distribution of data points. This method also tries to overcome the limitation of equal-width interval discretisation. In this research, all attributes of the dataset were discretised by this method.

3.3. Feature Selection

Feature selection was used to combat the curse of dimensionality and accelerate the training phase of machine learning algorithms. This was done by selecting only the most important or relevant features according to certain measures. Two significant classifications for feature selection are the wrapper and filter methods. Wrapper methods utilise the machine learning algorithm to test each feature subset. The result is typically better than filter methods but at the cost of further computational complexity. Filter methods are independent of the machine learning method to be applied but perform much faster. Data features were reduced in this process, but data integrity was also preserved to make it suitable for further analysis. Irrelevant and useless features were also eliminated for the quality preparation of data to obtain good results.

3.3.1. Information Gain

The concept behind information gain (also known as entropy) measures the bits of information available for class prediction. Given a single attribute, each value will be evaluated through Equation (6):

$$E(v) = -(P(2)\log_2 P(2) + P(1)\log_2 P(1) + P(0)\log_2 P(0)) \tag{6}$$

where $P(2)$ denotes the probability of class 2 occurring with the attribute value, $P(1)$ indicates the probability of class 1 occurrence, and $P(0)$ indicates the probability of class 0 occurrence. Given these values, the expected new entropy can be calculated through Equation (7):

$$E_{new}(v) = \sum P(v) * E(v) \tag{7}$$

where $P(v)$ denotes the probability of the value v occurring, and $E(v)$ indicates the entropy for this value. Then, the information gain using Equation (8) will be:

$$I(v) = E(v) - E_{new}(v) \tag{8}$$

Original entropy is simply the entropy using the probability of each target class occurrence. Given that the original entropy of the data remains static, the smaller the expected entropy value, the larger the information gain. In the context of feature selection, a feature with the lowest expected entropy will be seen as the most valuable by this measure.

3.3.2. Relief-F Algorithm

The Relief algorithm [19] is a generic method initially developed for classification problems with binary classes. It attempts to estimate the quality of predictors and of how well their values distinguish between instances near each other. For a randomly selected training instant R_i, the Relief algorithm finds its two nearest neighbours: one from the same class called the nearest hit H, and the other from the different class, called the nearest miss M. It updates the quality estimation $W[P]$ for all predictors P depending on their values for R_i, M, and H. If instances R_i and H have different values of the predictor P, then the predictor P separates two instances with the same class, which is not desirable, so the quality estimation $W[P]$ is decreased. On the other hand, if instances R_i and M have different values of the predictor P, then the predictor P separates two instances with different class values, which is desirable, so the quality estimation $W[P]$ is increased. The whole process was repeated m times. The Relief-F algorithm [20] is an improved version of the Relief algorithm used for classification problems with more than two classes. It employs more than a single nearest neighbour and can handle missing predictor values. The Relief-F algorithm is another extension to handle regression problems. In contrast to the majority of heuristic methods for estimating the quality of predictors, which assume the conditional independence of the predictors, relief algorithms can determine the quality of the predictors with high dependencies between themselves.

3.3.3. Wrapper Method

In the Wrapper method, a predictor (or classifier) is used to evaluate the feature subset. This method takes classifier performance, i.e., error rate, accuracy, etc., as a measure to determine the relative usefulness of a subset. Before the selection process is performed, we need to define the search space of all possible variable subsets and which classifier is used, and assess classifier performance and stopping criteria [21]. The subset search can be performed sequentially or heuristically, and the proposed subset is evaluated until maximum performance is gained with the minimum number of features. Since the Wrapper method uses particular classifiers as the main component for evaluation, the whole process highly relies on a specific classifier being used. The most popular classification algorithms used for the Wrapper method are SVM, RF, and ANN. Defining how to search the subset space is an important step in the Wrapper method. Generally, a subset search algorithm can be classified into two types: sequential selection algorithm (SSA) and heuristic search algorithm (HSA) [22].

The SSA technique can be performed in two ways: forward selection (SFS) and backward selection (BFS). Forward selection starts from an empty set of feature subsets, then adds a feature that maximises objective function one by one until there is no more improvement in objective function score. The subset that provides the best objective function score is chosen and validated. A backward selection has the same idea, but it starts from the full-feature set and removes the most features that reduced the objective function score. One drawback of SSA is that it is prone to "nesting effects", which means the already selected or removed feature cannot be removed or selected in later stages. Some variations of SSA are developed to avoid the nesting effect, such as "plus-L-minus-R" selection (LRS), sequential backward floating selection (SBFS), and sequential forward floating selection (SFFS). The HSA approach is based on heuristic optimisation using an evolutionary algorithm to find the optimal solution of the objective function. Genetic algorithm (GA) is often used for HSA. Individual features and output variables are represented as a gene. An individual represents a single solution containing possible feature combinations (in GA terms, also called chromosome). HSA tries to find an optimal solution by selecting the best individual in the population (collection of random solutions) and producing a possibly better set of solutions through mating, reproduction, and induced mutation [23].

The wrapper method can produce the best feature subset that suits a particular classifier and scores high in performance evaluation, typically better than the filter method. However, its reliance on particular classifiers and overtraining might lead to overfitting or poor generalisation. The wrapper method requires a training classifier model for each subset evaluation. An exhaustive search could result in the best accuracy but would be too expensive to perform, especially when the number of features or samples is enormous. Nevertheless, even with more advanced search algorithms, the computation required to achieve the desired criteria could still be too much.

3.3.4. Lasso

The famous least absolute shrinkage and selection operator (Lasso), proposed by Tibshirani [24], is very popular because of its variable selection property and has been used in many fields of statistics. This method shrinks values of some coefficients to zero by a constraint on the sum of absolute values of regression coefficients so that Lasso can serve as a tool for variable selection. The substantial difference between Lasso and the subset selection procedures or the information criteria is that Lasso selects variables, estimates the coefficients simultaneously, and retains good subset selection and ridge regression features. Lasso is a regularisation and variable selection algorithm that performs mostly better than other methods. Suppose we have a selected subset of features with size k, denoted by $\{s_1, s_2, \ldots, s_k\}$. $x_i = \left(x_i^{(s_1)}, x_i^{(s_2)}, \ldots, x_i^{(s_k)}\right)^T$ is the vector of selected features for individual i, and β_0 is the intercept, and $\beta = \left[\beta_{(s_1)}^T, \beta_{(s_2)}^T, \ldots, \beta_{(s_k)}^T\right]^T$ is the parameter vector. The simple logistic regression of the selected features is explained through Equation (9):

$$Pr(y_i = 1) = \frac{e^{\beta_0} + x_i^T \beta}{1 + e^{\beta_0} + x_i^T \beta} \tag{9}$$

We can estimate β by minimising the negative log-likelihood via Equation (10):

$$l(\beta_0, \beta) = -\frac{1}{n} \sum_{i=1}^{n} \left(y_i \left(\beta_0 + x_i^T \beta\right) - \log\left(1 + e^{\beta_0 + x_i^T \beta}\right)\right) \tag{10}$$

We add L1 Lasso (Least absolute shrinkage and selection operator) penalty for obtaining sparse solutions and enhancing predictive performance. The Lasso estimator is obtained from the penalised minus log-likelihood using Equation (11):

$$\hat{\beta}_{LASSO}(\lambda_1) = argmin_{\beta_0, \beta}\, l(\beta_0, \beta) + \lambda_1 ||\beta||_1 \tag{11}$$

where $||\beta||_1 = \sum_{j=1}^{p} \left|\beta_0\right|$, p is the total number of dummy variables of selected features, and λ_1 is the tuning parameter. Note that the intercept is not included in the penalty term. Lasso penalty corresponds to a Laplace before Bayesian inference. Hence, it will obtain a subset of important features with non-zero coefficients and shrink the reset to zero. Increasing λ_1 will shrink more coefficients to zero by adding a heavier penalty. Because this optimisation problem is convex, it can be solved efficiently for large data. There are several algorithms for calculating the Lasso estimator, among which coordinate descent performs the best. Coordinate descent optimises each parameter separately while holding all others fixed. Feature selection reduced the data dimensions by reducing the number of features. Initially, there were 108 attributes in our data set. Fourteen most-contributing attributes were selected for school performance measurement through various feature selection methods, as shown in Table 2.

Table 2. Significant feature selection through feature selection methods.

Information Gain	Wrapper Method	Relief-F	LASSO
School Gender	School Area	Total Schools	Total Computer Labs
Total Schools	Total Playgrounds	Total Teachers	Students without Furniture
Total Urban Schools	School Medium	Total Students	Total Rural Schools
Building Condition	Total Schools	Total Rural Schools	Total Urban Schools
Total Computer Labs	Total Urban Schools	Total Urban Schools	Open Air Class Sections
Total Available Classrooms	Classes	Open Air Class Sections	School Gender
Students without Furniture	Total Teachers	Total Computer Labs	Building Condition
Library Presence	Students with Furniture	Total Science Labs	Total Playgrounds
Open Air Class Sections	Total Students	Students with Furniture	Total Students
Students with Furniture	Students without Furniture	Students without Furniture	Deficiency of classrooms
Total Rural Schools	Open Air Class Sections	Total Playgrounds	Total Science Labs
Total Students	Total Available Classrooms	Total Available Classrooms	Total Teachers
Total Playgrounds	Building Condition	Deficiency of Classrooms	School Shift
School Shift	Total Science Labs	School Having Electricity Facility	Library Presence
Matric Result	Matric Result	Matric Result	Matric Result

3.4. Modelling

In this study, the following models were utilised for the performance measurement of institutions. Machine learning models are also widely used in the domain of healthcare [25–27], robotics [28,29], and business [30,31].

3.4.1. J48 Classifier

C4.5, known as J48, is a classifier first developed by Ross Quinlan and an extension of the ID3 algorithm. Most of the machine learning classifiers adopt greedy and top-down approaches for making a decision tree. In J48, classification is based on existing observations and training datasets; new data is labelled. While formulating a decision tree, the training dataset is partitioned into smaller partitions by dividing and conquering at each node. The dataset consists of collections of objects and objects that can be either an activity or an event. Each tuple of the dataset contains a class label that defines which object belongs to which class. If the tuples belong to different classes, then further splitting can be performed. While partitioning a dataset, a heuristic approach is followed, which chooses an attribute for the best partition known as the selection measure. The type of branching formation at each node is the responsibility of this selection measure. Like information gain, the Gini index is an example of partitioning the node to multi-label and binary, respectively [32]. For a better working understanding, let us have dataset $S = X_1, \ldots, n, C_i$, where C_i denotes the dependent variable n representing the number of independent variables, the value of i can be from 1, 2, : :: , K. K represents the classes of the dependent variable. At every partition, a new node is added to the decision tree. In S partition, X is chosen for further partitioning into different sets like $S_1, S_2, \ldots, S_l$. These new child nodes are then added into the main node S of the decision tree. The primary node S is labelled with text X and newly created partitions $S_1, S_2, \ldots, S_l$ are partitioned again recursively. The partition will not be further split into sub-partitions if all records in a partition have identical class labels. Its corresponding leaf will be labelled as a dependent variable.

The following steps are followed to construct a decision tree using J48. In step 1, we calculate the Entropy of training set S through Equation (12).

$$Entropy(S) = -\sum_{i=1}^{K}\left\{\left[\frac{freq(C_i, S)}{|S|}\right] log_2\left[\frac{freq(C_i, S)}{|S|}\right]\right\} \tag{12}$$

where samples in the training set are represented with $|S|$. C_i is identified as dependent variable, $i = 1, 2, : : : , K$. K represents classes belong to the dependent variable, and freq(C_i, S) has total samples that class C_i contains.

In step 2, for partition, Information Gain $X(S)$ is calculated for the test attribute X as explained in Equation (13):

$$Information\ Gain_x(S) = Entropy(S) - \sum_{i=1}^{L} \left[\left(\frac{|S_i|}{|S|} \right) Entropy(S_i) \right] \tag{13}$$

where S_i is denoted as a subset of S for that particular ith output, and $|Si|$ defines the dependent variables of a subset S_i. L represents the test outputs, X. That subset will be selected as a threshold for a specific attribute partition to provide maximum information gain. S and S-Si partition will be the branch of the node. If the instance belongs to the same class, then the tree's leaf will be labelled and returned as a dependent variable (class).

In step 3, partition information value *Split Info(X)* will be calculated by acquiring for S partitioned into L subsets through Equation (14):

$$Split\ Info(X) = -\sum_{i=1}^{L} \left[\left(\frac{|S_i|}{|S|} \right) log_2 \left(\frac{|S_i|}{|S|} \right) + \left(1 - \left(\frac{|S_i|}{|S|} \right) \right) log_2 \left(1 - \left(\frac{|S_i|}{|S|} \right) \right) \right] \tag{14}$$

In step 4, we calculate *Gain Ratio(X)* using Equation (15):

$$Gain\ Ratio(X) = \frac{Information\ Gain_x(S)}{Split\ Info(X)} \tag{15}$$

In step 5, based on the value of the gain ratio, the attribute having the highest value is declared root node, and the same computation is repeated from step 1 to step 4 for intermediate nodes till all the instances are exhausted and reach the leaf node as per step 2 [33].

3.4.2. Support Vector Machines

Support vector machines (SVMs) are primarily constructed for multiclass classification, although they can perform binary separation. The idea of SVMs is that a classification problem with N number of input features can be solved by finding a hyperplane of dimension $N - 1$. The hyperplane separates the N-dimensional space in N parts where the data points in the same subspace also belong to the same class. The equation for the separating hyperplane will have several solutions [34]. For the sake of simplicity, consider a linear SVM where $N = 3$, then the hyperplane is a line. The line can be moved sideways between its two closest points to separate and even be tilted in new angles and still separate training data points of the N-classes into their own spaces. A poorly chosen hyperplane out of the alternatives may make the performance on test data suffer, although the training performance is the same. A similar problem will be found in higher dimensions and non-linear settings as well. To obtain a good model, a good hyperplane must be found. One such hyperplane is the maximum margin hyperplane. The maximum margin hyperplane is the maximum distance to the data points closest to the hyperplane, thus a hyperplane with the maximal possible margin.

The data points on the margin to the hyperplane are called "support vectors" since they support the placement of the hyperplane. The maximal margin hyperplane is only dependent on the support vectors for its positioning. If the training set is changed by adding or removing data points, it will not affect the classifier unless the set of support vectors is altered. However, it is not satisfactory that the classifier can be fundamentally changed by adding just one training sample. Moreover, the scenario that there is no perfectly separating hyperplane needs a solution. A solution for both problems is to introduce a soft margin. The soft margin introduces an error tolerance of the model, which allows some of the training data points to be on the wrong side of the margin, or even the wrong side of the hyperplane. The constraint added is that the total errors may sum up to a specific constant but no more. Data points within the margin will also be considered as support vectors.

A linear separating hyperplane will follow Equation (16):

$$0 = \beta_0 + \beta_1 X_1 + \beta_2 X_2 + \ldots + \beta_N X_n \tag{16}$$

where $\beta_i \in \{\beta_i, \ldots, \beta_N a$ are the parameters to find by training. The corresponding classification function is explained in Equation (17):

$$f(x) = \beta_0 + \beta_1 x_1 + \beta_2 x_2 + \ldots + \beta_N x_n \tag{17}$$

Here, x_i are the features of the sample to classify. The class of a new data point is determined by whether $f(x)$ has a value above or below zero. The linear classification function of the SVM utilises the inner products of the observations. Therefore, the classification function can be rewritten as described in Equation (18):

$$f(x) = \beta_0 + \sum_{i=1}^{N} \alpha_i K(x, x_i) \tag{18}$$

where α_i are the parameters found by the training, and $K(x, x_i)$ is the inner product between observations (Equation (19)):

$$K\left(x_i, x'_i\right) = \sum_{j=1}^{p} x_i x'_i \tag{19}$$

SVM has primarily been constructed for multiclass classification. The idea of SVM is that a classification problem with N number of input features can be solved by finding a hyperplane of dimension $N - 1$. SVM classification is an optimisation problem. Linear discriminant analysis (LDA) and kernel functions are two analytical solutions used for optimisation. We utilised kernel methods for SVM to transform a linear classifier into a non-linear classifier. In the linear classifier, the inner product is called the kernel function, or Kernel for short. The kernel of a classifier quantifies the similarity of two observations. To separate classes that have non-linear boundaries, the hyperplane must be described by a non-linear equation. If the non-linear equation is polynomial, the classifier function will use a polynomial kernel, where d is the degree presented in Equation (20):

$$K\left(x_i, x'_i\right) = [1 + \sum_{j=1}^{p} x_i, x'_i]^{d} \tag{20}$$

Past researchers also utilised LDA, which uses the entire dataset to estimate covariance matrices and is also prone to outliers that are a significant limitation; hence, we utilised kernel functions instead of LDA. Our dataset also had diversity in values, performance or percentage of results that differed significantly between schools in big cities such as Lahore and the schools in Southern Punjab, which made the performance values of the schools in backward areas an outlier. As in our dataset, few attributes or features had outlier values, e.g., some negative values were observed in the "students without furniture" attribute, which was replaced with zero after comparing and analysing the rest of the instances. LDA does not work well if the dataset is imbalanced (i.e., the number of objects in various classes is different). Our dataset had three classes in the class label that were different because only a few cases were good and bad, whereas most cases were in the medium category. We implemented the LDA, but the results were not persuasive at all. We chose SVM for further experiments and analysis.

3.4.3. Random Forest

The random forest (RF) algorithm is based on the decision tree model and is straightforward, flexible, and fast. RF, including nominal, binary can handle different types of data and numerical, and has high predictive capability. RF works by building multiple trees and aggregating trees to generate efficient results [35]. The trees are generated based on seeds. Randomness in seeds generates random trees that are efficient and result in better prediction.

Similarly, a random and different subset of attributes gives more accurate results on large datasets. The classifications for the new input data are based on each contributed tree's functions for one class. RF then performs prediction by checking the plurality of votes for the new instances. Every internal node tests an attribute, and the test result is represented by edge [36]. RF is based upon the concept of bagging and boosting. In bagging (bootstrapping), a model is constructed again and again, sampling from a large set of examples used for training, and then results are aggregated through a majority vote. So, to construct a good classifier from uncorrelated weak classifiers, boosting is an optimal solution. The tuning of hyperparameters controls the number of features in each tree [37].

RFs are ensemble learners, which means many weak "base learners" contribute their votes for prediction. Base learners are called decision trees in RF, consisting of a branching composite of binary decisions for separating the data into classes. At each node of the tree, the input is separated by choosing threshold t and a single feature d. The resulting split should have minimal impurity (by mean of class labels). Entropy H is presented for two-class learning as explained in Equation (21):

$$H = - \sum_{c=1}^{2} \hat{\pi}_c log\ log\ \hat{\pi}_c \tag{21}$$

where c denotes the class and $\hat{\pi}_c$ represents the proportion to the examples in c. Maximisation in information gain is equivalent to minimising entropy. In RF, the number of trees and selection of features are controlled by tuning hyperparameters. An importance matric can be assigned to the features based on their impact on node impurity, weighted by the importance and worth of the node in classification [38]. In a single tree t, this feature importance $I_{d,t}$ for feature d is formulated as presented in Equation (22):

$$I_{d,t} = \sum_{n \in N_d} \left[\left(H_{pre,n} - \sum_{s=1}^{2} H_{post}, n, s \right) \times P_n \right] \tag{22}$$

H_{pre}, denotes the entropy before node splitting, and H_{post}, n, s represents the entropy after the split of child node s. N_d represents the set of all nodes split by feature d. For the given node n, P_n denotes the proportion of samples at that node. The $I_{d,t}$ scores are overall averages of the N_t built decision tree T for preparation of a resultant I_d importance weighting: $I_d = \frac{1}{N_t} \sum_{t \in T} I_{d,t}$ [39].

Combining the multiple decision trees to attain better variance reduction results is also important, but there is a potential downside. The RF algorithm selects a fixed number of predicators from the available features in the pool at each split to overcome this. The predicators of all individual decision trees are combined to prepare the final predicator by averaging the majority vote [40]. A few more reasons support the excellent prediction power of the RF algorithm and its wide adoption. One key feature of this algorithm is its stabilisation with fewer iterations than another state-of-the-art ensemble method such as boosting. Secondly, it is working, visualising, and tuning on different inputs that influence and attract users.

3.4.4. Rotation Forest

The main difference from other tree algorithms is that rotation forest does not require as many trees to be created to achieve impressive accuracy. Unlike the random forest, the rotation forest is used when the number of ensembles is small. Rotation forest is an ensemble-based method first proposed by Rodriguez et al. [41]. The rotation forest model requires some parameters that the user defines. Hence, it spares much time in creating trees, which is relatively time-consuming.

Interestingly, the authors of the algorithm claim that an underlying estimator can be not only a tree but anything else, as well, although what remains unchanged is that it still uses bagging as one of the basic techniques. The user should specify the number of trees. When that is done, the algorithm looks like this:

For each tree T, perform the following:

1. Split the attributes in training set into K non-overlapping subsets of equal size.
2. For each of the K datasets with k attributes, perform the next steps.
3. Create a rotation matrix of size $N \times N$, where N is the total number of attributes. In the matrix, each principal component should match the position of the feature in the original training dataset.
4. Project the training dataset on the rotation matrix using matrix multiplication.
5. Build a decision tree with the projected dataset.
6. Store the tree and rotation matrix.

3.4.5. Artificial Neural Networks

An artificial neural network (ANN) consists of several interconnected processing units that process information. It contains three types of layers: the first is called the input layer, then the hidden layer, and the last output layer. The transformation is carried out through the centred layer (hidden layer) between the input and output layer through units to detect complex patterns and learns accordingly. The idea of ANN working has been perceived by the working mechanism of the human brain. The brain consists of billions of neurons, and a single neuron is known as a perceptron, and each neuron is connected to others by axons. The neurons are finally connected with the synapses, which allow neurons to pass the signal. The neural network is formed with a large number of simulated neurons.

Similarly, ANN contains multiple nodes in itself that are connected. The joining among units is denoted by weight. Inputs passed to the ANN consist of different values that are connected with weight vectors. The weight can be either positive or negative. For results generation, the function used to sum the weights and map to output is $y = w_1x_1 + w_2x_2$.

ANN has been used for both supervised and unsupervised learning. This study applied supervised learning because the input and output were known and provided to the model. The model was tuned with different values to adjust the weights to the best to obtain the expected efficient output [35]. In multiclass classification, classifiers are used to predict multiple outcomes. In this study, a multiclass neural network was used to build a classification framework. Let us have K classes and want to classify one instance from one class. Then, the best choice is to use a linear neural network with multiclass classification. It is an extension of the binary classification setup. The second layer node will generate output as *0, 1 ... K-1*. The basic working principle of a multiclass artificial neural network is shown in Figure 2.

We have $|w| = MK$, where M denotes the number of features and K represents classes. If $K = 3$ and $M = 3$, then the total weight will be formed as 9. To support the neural network view of multinomial logistic regression, we receive help from binary logistic regression (Equation (23)) as:

$$P(Y = y | X = x_i, w) = \frac{1}{1 + exp(-yw^T x_i)} \tag{23}$$

In the case of K number of classes, we will have Equation (24):

$$P(Y = k | X = x_i, w) = \frac{\exp\left(w_k^T x_i\right)}{\sum_{k'}^{K} \exp\left(w_{k'}^T x_i\right)} \tag{24}$$

In the above equation, Y is the dependent variable representing a value we are trying to predict. The variables *(Xi = 1 to n)* are used to predict values for the dependent variable. W represents weight value, one for each data instance. It shows the strength and type of relationship with a particular data instance with Y. Larger values of weight represent a stronger relationship [42].

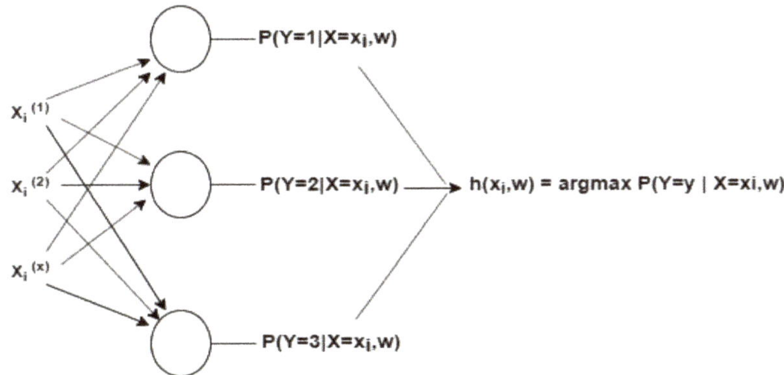

Figure 2. The basic working mechanism of a multiclass artificial neural network.

4. Results

In machine learning classification, the results are measured on the basis of accuracy, recall, precision, F-measure, ROC, and root mean square error (RMSE). Accuracy is the ratio of correct predictions of the sample over the total number of predictions. The results may vary among DM models due to internal changes in processing functionality. All of the evaluation metrics are built on four types of classifications: true positives (*TP*), true negatives (*TN*), false positives (*FP*), and false negatives (*FN*).

$$Accuracy = \frac{No.\ of\ correct\ predictions}{Total\ No.\ of\ predictions} \tag{25}$$

For binary classification, the accuracy is measured using Equation (25) or Equation (26):

$$Accuracy = \frac{TP + TN}{TP + TN + FP + FN} \tag{26}$$

TP represents true positive, *TN* is a true negative, *FP* is false positive, and *FN* represents false negative. *AUC-ROC* is also used to calculate the performance of multi-classification problems. *ROC* is a probability curve that stands for Receiver Operating Characteristics, and Area under the Curve (*AUC*) measures the degree of separability. It states the capability of the model to distinguish between classes, and the higher the *AUC*, the better the model distinguishes between classes. The *ROC* curve is plotted with *TPR* against *FPR*, where *TPR* is on the *y*-axis and *FPR* is on the *x*-axis. True positive rate (*TPR*) or recall value is calculated through Equation (27):

$$Recall\ (TPR) = \frac{TP}{TP + FN} \tag{27}$$

False positive rate (*FPR*) is calculated through Equation (28):

$$FPR = \frac{FP}{FP + TN} \tag{28}$$

Precision is used to determine the number of predicted positive instances correctly classified by the algorithm as presented in Equation (29).

$$Precision = \frac{TP}{TP + FP} \tag{29}$$

F-measure is used to represent the harmonic mean between two parameters, precision and recall, as shown in Equation (30). A high value of F-measure indicates that both precision and recall are reasonably high.

$$F - measure = \frac{2 * (Recall * Precision)}{Recall + Precision} \tag{30}$$

$RMSE$ is a frequently used measure of the differences between values predicted by a model and observed values. The $RMSE$ represents the sample standard deviation of the differences between predicted values and observed values. where y'_i is the predicted value and y_i is the true value for subject i. The $RMSE$ values are calculated through Equation (31):

$$RMSE = \sqrt{\frac{\sum_{i=1}^{n} \left(y'_i - y_i\right)^2}{n}} \tag{31}$$

As ML models grow rapidly, they also need more tuning and configuration. This tuning often comes in the form of hyperparameters. The hyperparameter describes all parameters that have to be determined before the actual process of fitting a model to the data is started. These hyperparameters exist because data-based models are designed to work in different scenarios, requiring both algorithm and model modifications. In the past, these modifications were often performed by using domain knowledge or rules of thumb. However, hyperparameters are generally challenging to set. Hyperparameters in machine learning describe variables that modify how a particular model is derived from data. These parameters can modify the algorithm that performs this process, but they can also be a model parameter that the algorithm cannot reasonably determine. Most model parameters are determined through training by applying the machine learning algorithm to the data. Hyperparameters are usually not independent of each other. The number of possible combinations of hyperparameters increases exponentially with the number of hyperparameters. Because training machine learning models is computationally expensive, the main goal is to find good or optimal points with as few function evaluations as possible. A common hyperparameter in the neural network case is the learning rate. It changes the rate at which neuron weights are adjusted per learning step and is essential for the performance of a neural network. While the consensus is that low learning rates slow learning down, high learning rates might keep the network from converging. This study utilised J48, SVM, RF, rotation forest, and ANN for training with various hyperparameters. The ranges of the hyperparameters are presented in Table 3.

Table 3. The ranges of the hyperparameters of the classifiers.

Classifier	Hyperparameters		
J48	Confidence factor [0.05–0.50]	Minimum number of instances per leaf nodes [2–6]	Random seed [1]
SVM	Kernel type [1–3]	Epsilon [1.0×10^{-12}]	Random seed [1]
RF	Number of trees [50,100]	Maximum depth of trees [15]	Random seed [7–11]
Rotation Forest	Ensemble size [5–15]	Maximum depth of trees [15]	Random seed [1]
ANN	Learning rate [0.3]	Number of hidden layers [2]	Random seed [6–20]

Usually, machine learning models split data sets into training and testing sets. Training is used to train the model while testing sets are used to test the model. Various approaches such as k-fold cross-validation and train test Split are used to validate results [8,35,43]. In the train test split, values are set for the model on how much data the model has to train and test. Mostly, it performs well for large datasets. In this research, the 10-fold cross-validation technique was utilised to obtain effective results on small datasets. The cross-validation approach works, as it splits the dataset into three portions to train, test, and validate the set. K sets the value to guide the model regarding how many equal folds of datasets to prepare after division. The first fold was used for testing purposes, the remaining $k - 1$ folds were used to train the model, and the whole process was repeated k times.

In this study, various machine learning models with feature selection methods were used. The stopping criteria for feature selection methods was set so that when the performance of the models decreased, the execution of feature selection methods stopped. The fourteen most significant features, as selected through feature selection methods, were used. The J48 algorithm derived results by using the approach of post-pruning. Post pruning is the process of evaluating decision tree error at each decision tree junction. The pruning of decision trees optimises the computational efficiency of the model. The pruning method reduces the size of the tree and unnecessary complexity. To test the effectiveness of post-pruning, the hyperparameter is often labelled as a confidence factor. If the value of the confidence factor is kept low, then the amount of post-pruning is decreased.

Moreover, the minimum instances per leaf node are set, which means to set the minimum amount of separation. It guarantees that at least two of the branches have the minimum number of instances at each split. For example, if one instance is separated from 100 instances, it does not give much information. The J48 decision tree model was combined with feature selection methods. The best results were obtained using the relief-F-based feature selection technique, which achieved maximum accuracy of 68.5% with an ROC value of 0.63 when the model has was with a confidence factor of 0.50. The minimum number of instances per leaf node was 6. The complete results for the J48 classifier with feature selection methods are presented in Table 4. After the J48 decision tree, SVM was utilised to obtain more effective results.

In SVM, the kernel type was selected. Hence, the kernel type, i.e., PolyKernel, Normalized PolyKernel, and radial basis function (RBF) Kernel, were chosen for better performance. In this experiment, the model was initially tuned with PolyKernel, and the predicted results revealed an accuracy of 61.4% while employing an information gain-based feature selection method. The model was again tuned with different feature selection methods, and this process was repeated multiple times until the highest accuracy value was achieved. The highest accuracy of 68.5% with RBF Kernel was achieved while employing the Relief-F based feature selection method. Complete results for the SVM classifier with various feature selection methods are presented in Table 5. After the SVM model, the random forest was utilised to obtain more effective results. Several machine learning models such as random forest and ANN are non-deterministic, requiring a random seed argument for reproducible results. Random seed denotes the random initial value for the algorithms.

We used different seed values to perform the experiments. In random forest, multiple trees were built with seeds that made a forest because the similar nature of trees decreased model performance. So, to achieve better performance, individual trees were built differently. The randomness in the generation of trees could be achieved with the use of random seeds. In this experiment, the model was initially tuned with 50 trees with random seed =7, and the predicted results revealed an accuracy of 68.5%. In contrast, the information gain-based feature selection method was employed. The model was again tuned with different combinations of values, and this process was repeated multiple times until the highest accuracy value was achieved. The highest accuracy of 71.3% with an ROC of 0.65 was obtained when the number of trees was set to 100, and the random seed was set to 8 while using the relief-F-based feature selection method. Complete results for the RF classifier with various feature selection methods are presented in Table 6. After the random forest, the rotation forest was utilised to obtain more effective results. In the rotation forest, ensemble trees were built because the similar nature of trees decreased the model performance. So, to achieve better performance, individual trees were constructed differently. In this experiment, the model was initially tuned with an ensemble size of 5, and the predicted results revealed an accuracy of 65.2%.

Table 4. Performance of J48 classifier through various feature selection techniques.

Feature Selection Method	Confidence Factor	Minimum Number of Instances Per leaf Nodes	Accuracy (%)	Precision (%)	Recall (%)	ROC	RMSE	F-Measure			
								Low	Medium	High	Weighted Average
Information Gain	0.05	2	65.0	0.53	0.65	0.51	0.42	0.23	0.67	0.35	0.56
	0.25	2	62.9	0.61	0.62	0.60	0.47	0.26	0.76	0.39	0.62
	0.50	**6**	**67.1**	**0.64**	**0.67**	**0.63**	**0.41**	**0.28**	**0.79**	**0.42**	**0.65**
Wrapper Method	**0.05**	**2**	**62.7**	**0.62**	**1.00**	**0.50**	**0.43**	**0.25**	**0.88**	**0.41**	**0.77**
	0.25	2	60.4	0.56	0.60	0.47	0.50	0.18	0.69	0.34	0.56
	0.50	6	61.7	0.55	0.61	0.61	0.46	0.19	0.68	0.33	0.57
Relief-F	0.05	2	65.7	0.58	0.65	0.57	0.43	0.11	0.81	0.27	0.61
	0.25	2	63.6	0.60	0.63	0.58	0.45	0.22	0.79	0.27	0.61
	0.50	**6**	**68.5**	**0.68**	**0.68**	**0.63**	**0.40**	**0.35**	**0.80**	**0.50**	**0.68**
LASSO	0.05	2	55.4	0.51	0.55	0.50	0.39	0.23	0.62	0.31	0.52
	0.25	**2**	**59.2**	**0.53**	**0.59**	**0.52**	**0.41**	**0.18**	**0.66**	**0.33**	**0.57**
	0.50	6	51.7	0.47	0.51	0.46	0.35	0.26	0.59	0.28	0.49

Table 5. Performance of SVM classifier through various feature selection methods.

Feature Selection Method	Kernel Type	Accuracy (%)	Precision (%)	Recall (%)	ROC	RMSE	F-Measure			
							Low	Medium	High	Weighted Average
Information Gain	PolyKernel	61.4	0.61	1	0.50	0.46	0.19	0.86	0.35	0.75
	Normalized PolyKernel	**62.7**	**0.62**	**1.00**	**0.50**	**0.45**	**0.23**	**0.88**	**0.32**	**0.77**
	RBF Kernel	62.0	0.62	1.00	0.50	0.46	0.21	0.87	0.35	0.77
Wrapper Method	PolyKernel	60.9	0.61	1.00	0.50	0.46	0.20	0.84	0.37	0.75
	Normalized PolyKernel	58.9	0.59	1.00	0.50	0.47	0.18	0.86	0.35	0.74
	RBFKernel	**63.6**	**0.63**	**1.00**	**0.50**	**0.45**	**0.22**	**0.87**	**0.33**	**0.77**
Relief-F	**PolyKernel**	**68.5**	**0.68**	**1.00**	**0.48**	**0.41**	**0.25**	**0.92**	**0.39**	**0.81**
	Normalized PolyKernel	67.8	0.68	0.99	0.49	0.42	0.27	0.90	0.34	0.80
	RBFKernel	**68.5**	**0.68**	**1.00**	**0.48**	**0.41**	**0.21**	**0.90**	**0.42**	**0.81**
LASSO	PolyKernel	53.9	0.53	0.98	0.47	0.43	0.19	0.75	0.29	0.65
	Normalized PolyKernel	**56.5**	**0.56**	**1.00**	**0.50**	**0.45**	**0.23**	**0.81**	**0.34**	**0.71**
	RBFKernel	55.2	0.55	0.96	0.49	0.44	0.20	0.68	0.31	0.67

Table 6. Performance of random forest classifier through various feature selection techniques.

Feature Selection Method	Number of Trees	Random Seed	Accuracy (%)	Precision (%)	Recall (%)	ROC	RMSE	F-Measure			
								Low	Medium	High	Weighted Average
Information Gain	50	7	**68.5**	**0.60**	**0.68**	**0.64**	**0.39**	**0.21**	**0.83**	**0.16**	**0.62**
	100	8	67.1	0.56	0.67	0.66	0.39	0.23	0.78	0.18	0.59
	100	11	66.4	0.57	0.66	0.65	0.39	0.15	0.78	0.21	0.59
Wrapper Method	50	7	**65.9**	**0.56**	**0.65**	**0.66**	**0.41**	**0.24**	**0.75**	**0.19**	**0.57**
	100	8	63.6	0.50	0.63	0.66	0.42	0.16	0.71	0.22	0.55
	100	11	63.6	0.49	0.63	0.65	0.41	0.21	0.69	0.17	0.54
Relief-F	50	7	70.6	0.64	0.70	0.63	0.39	0.18	0.84	0.21	0.63
	100	8	**71.3**	**0.65**	**0.71**	**0.65**	**0.39**	**0.25**	**0.83**	**0.16**	**0.64**
	100	10	69.2	0.60	0.69	0.64	0.39	0.17	0.83	0.14	0.62
LASSO	50	7	53.9	0.44	0.53	0.51	0.41	0.23	0.64	0.19	0.45
	100	8	55.4	0.47	0.55	0.53	0.44	0.26	0.65	0.17	0.48
	100	**10**	**57.5**	**0.49**	**0.57**	**0.49**	**0.46**	**0.24**	**0.68**	**0.23**	**0.50**

In contrast, the information gain-based feature selection method was employed. The model was again tuned with different combinations of values, and this process was repeated multiple times until the highest accuracy value was achieved. The highest accuracy of 73.2% was obtained when the ensemble size was set to 15 while using the relief-F based feature selection method. Complete results for the rotation forest classifier with various feature selection methods are presented in Table 7.

After J48, SVM, random forest, and rotation forest, artificial neural networks were used to achieve more efficient results. The model was used to create a neural network that predicted the target based on multiple input values. The model was tuned with different parameters to generate the best result. One of them is known as random number seed. It was used to ensure repeatability across runs of the same experiment. The model was initially tuned with 6 random seeds, and the predicted accuracy was recorded as 79.0% while using the information gain-based feature selection method. It was observed that the model predicted its best results with an accuracy of 82.9% when the number of random seeds was 10 while utilising the relief-F based feature selection method. Complete results for the artificial neural network classifier with different feature selection methods are presented in Table 8.

Among the five classifiers utilised in this study, the artificial neural network outperformed and obtained the highest accuracy of 82.9% while utilising the relief-F based feature selection technique, as shown in Figure 3. It was observed that ANN also performed efficiently while utilising other feature selection techniques. The performance of ANN was also good while evaluating other performance metrics.

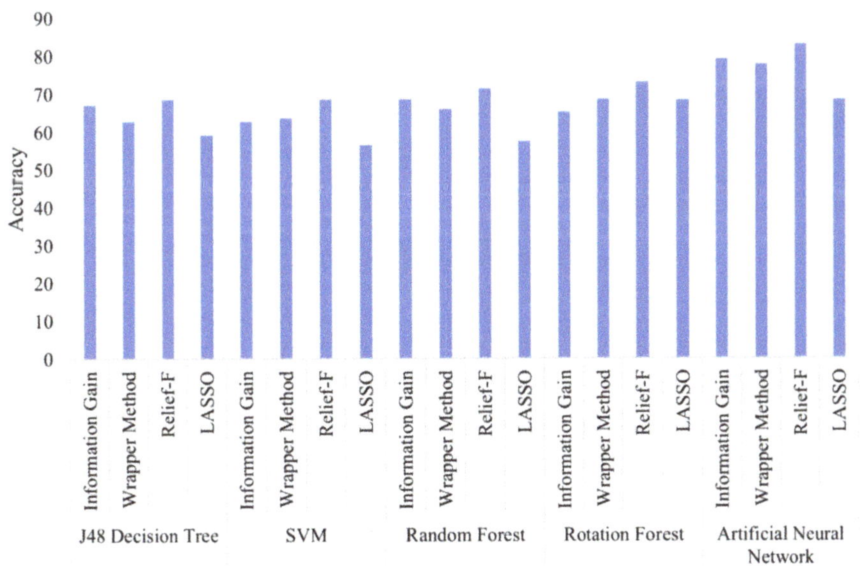

Figure 3. Performance comparison of models based on accuracy with utilisation of various feature selection techniques.

Table 7. Performance of rotation forest through various feature selection techniques.

Feature Selection Method	Ensemble Size	Accuracy (%)	Precision (%)	Recall (%)	ROC	RMSE	F-Measure			
							Low	Medium	High	Weighted Average
Information Gain	5	61.0	0.54	0.60	0.60	0.45	0.18	0.67	0.31	0.56
	10	62.5	0.60	0.61	0.59	0.46	0.25	0.74	0.28	0.61
	15	**65.2**	**0.59**	**0.64**	**0.56**	**0.42**	**0.19**	**0.72**	**0.30**	**0.60**
Wrapper Method	5	66.0	0.56	0.65	0.64	0.38	0.23	0.71	0.26	0.58
	10	67.4	0.55	0.66	0.65	0.38	0.25	0.77	0.23	0.58
	15	**68.5**	**0.52**	**0.52**	**0.70**	**0.40**	**0.19**	**0.63**	**0.21**	**0.52**
Relief-F	5	68.9	0.67	0.99	0.47	0.40	0.24	0.93	0.35	0.80
	10	70.5	0.62	0.69	0.62	0.38	0.25	0.72	0.31	0.62
	15	**73.0**	**0.69**	**0.90**	**0.78**	**0.38**	**0.26**	**0.95**	**0.37**	**0.81**
LASSO	**5**	**61.3**	**0.54**	**0.61**	**0.64**	**0.43**	**0.21**	**0.66**	**0.28**	**0.55**
	10	58.4	0.51	0.58	0.52	0.40	0.18	0.63	0.25	0.52
	15	60.7	0.53	0.60	0.56	0.42	0.19	0.64	0.26	0.54

Table 8. Performance of multiclass neural network through various feature selection techniques.

Feature Selection Method	Random Number of Seeds	Accuracy (%)	Precision (%)	Recall (%)	ROC	RMSE	F-Measure			
							Low	Medium	High	Weighted Average
Information Gain	**6**	**79.0**	**0.68**	**0.68**	**0.80**	**0.30**	**0.60**	**0.95**	**0.75**	**0.92**
	10	78.0	0.67	0.67	0.78	0.32	0.18	0.75	0.34	0.66
	20	73.4	0.60	0.60	0.78	0.38	0.16	0.71	0.32	0.60
Wrapper Method	6	77.1	0.65	0.65	0.79	0.35	0.18	0.74	0.34	0.65
	10	68.7	0.53	0.53	0.70	0.40	0.22	0.66	0.28	0.53
	20	**77.6**	**0.66**	**0.66**	**0.79**	**0.35**	**0.19**	**0.75**	**0.33**	**0.66**
Relief-F	6	77.1	0.66	0.66	0.79	0.35	0.17	0.73	0.37	0.66
	10	**82.9**	**0.68**	**0.68**	**0.84**	**0.27**	**0.62**	**0.95**	**0.90**	**0.94**
	20	77.6	0.66	0.66	0.79	0.35	0.24	0.72	0.39	0.66
LASSO	**6**	**68.3**	**0.57**	**0.68**	**0.76**	**0.29**	**0.23**	**0.77**	**0.33**	**0.67**
	10	65.6	0.54	0.65	0.73	0.27	0.21	0.66	0.30	0.56
	20	62.5	0.53	0.62	0.71	0.24	0.19	0.63	0.28	0.53

5. Discussion

In the past, most educational research has been focused on and evaluated students' academic performance in specific institutions or regions. Performance was calculated with consideration for various influences including socio-economic and demographic factors as well as students' personal, family, and academic backgrounds. Apart from student academic results, other factors also have a substantial impact on the performance of any educational institution. The present study focused on the importance of other highly influential factors along with student academic results, such as the students per teacher ratio, the number of schools in a region, whether schools were located in rural or urban areas, the availability or lack of classrooms, electrical facilities in schools, availability or lack of furniture for students, open-air classes, computer lab facilities, science labs, and playgrounds in schools. Previous research [44–48] suggested that data pre-processing (normalisation, discretisation) techniques enhanced classifier performance, as these techniques reduce the biases among features. Furthermore, related studies showed that the min-max normalisation method performed better than other data normalisation methods [49–51]. It has also been observed in related studies that binning-based data discretisation techniques outperformed other techniques based on their results [52–54]. This study will help in the identification of underperforming regions based on institutional performance. It will also support governance in performance monitoring, policy formulation, target-setting, evaluation, and reforms to address the issues and challenges of education. In this research, various feature selection methods were combined with machine learning models to obtain efficient results. The fourteen most significant features were used, as selected through feature selection methods. The J48 decision tree model was combined with feature selection methods. The best results were obtained using the relief-F-based feature selection technique, which achieved maximum accuracy of 68.5% with an ROC value of 0.63. The highest accuracy of 68.5% was achieved with the SVM (RBF kernel) model while employing the relief-F based feature selection method. After the SVM model, the random forest was utilised to obtain more effective results. The highest accuracy of 71.3% with an ROC of 0.65 was obtained. After the random forest, the rotation forest was utilised to obtain more effective results. The highest accuracy of 73.2% was obtained. After J48, SVM, random forest, and rotation forest, artificial neural networks were used to achieve more efficient results. It has been observed that this model predicted the best results with an accuracy of 82.9% while utilising the relief-F based feature selection method. The artificial neural network outperformed and yielded the highest accuracy, of 82.9%, among the five classifiers employed in this study. The performance of ANN also proved efficient while evaluating other performance metrics. It was also observed that the target class (medium) results were better than other target classes (low and high). This is because the number of instances in the medium class were significantly higher than in the high and low classes. The performance of machine learning models is better when trained on large datasets. In our study, the performance of machine learning models on medium classes was also high due to the large amount of data as compared to other classes.

This study provides additional support for researchers to employ the ANN model and apply it to social science studies. Moreover, this study showed that there is value in including special education-related predictors to improve classification accuracy. The study demonstrated how geographical and demographic variables could all add to the classification accuracy of prediction models. Lastly, the study results offered strong evidence that school facilities are highly predictive for the performance measurement of public schools. Classification into high, medium, and low support levels could also help to illustrate the relationship between variables and classification levels. More importantly, it could highlight the importance of going beyond single-variable, single-threshold early warning systems (e.g., systems that focus on only one KPI), which overlook complex interactions among predictors. One variable is not sufficient to predict measurements of public school performance. The proposed model based on ANN produces more accurate prediction values than the other existing approaches because of its heuristic learning and correction technique. The proposed work was developed on the basis of a bio inspirational approach for increasing the performance of the prediction process. ANN assigns weights based on trial and error during the training phase. This proposed work utilised the knowledge of the genetic algorithm to assign the weights of the hidden nodes, and thus its expected outcome and the actual outcome were closely matched. Hence, the proposed model's error rate is very low compared to other algorithms, while its prediction accuracy is also greatly improved.

On the map of the world, Pakistan is facing severe social, demographic, and educational disparities. It is ranked 143rd out of 144 countries on the Global Gender Gap (GGG) index with a score of 0.546, the worst in South Asia [55]. Among South Asian counties, Pakistan's performance in education is not reasonably satisfactory. Moreover, its educational disparities are higher, and significant efforts towards alleviating them have not been observed. Pakistan consists of five provinces, of which Punjab is the most populous. Punjab accounts for more than 56 percent of Pakistan's total population and 52 percent of its gross domestic product. Punjab consists of nine divisions and 36 districts. In Punjab, demographic disparities exist among the various districts [56]. Lahore (its developed district) ranks first and Muzaffargarh (underdeveloped district) last on the Human Development Index. In terms of educational disparities measured in average years of schooling, Muzaffargarh is more deprived, with 4.41 years for males and 1.95 for females, contrary to Lahore, with an average of 8.5 years of education for males and 7.34 years for females.

The same trend is found in all other provinces [57]. One of the probable reasons might be the strong family system in Pakistan, which places all economic responsibility on males, whereas females are not supposed to earn or spend within the family. Hence, education, whose primary purpose is to help secure jobs and livelihoods, might be male-focused. In addition, cultural values in Pakistan do not support the unrestricted mobility of females. They must be accompanied by male members of their families when travelling. Thus, the preferences for educating females are lower within a family. Such values are stronger in rural areas, where education appears to be considered a luxury for girls. Consequently, many females discontinue their education after exhausting the available resources in their hometowns, leading to educational disparities.

The Annual Status of Education Report: Pakistan (ASER-PAK) 2018 presented the current education status in Pakistan in all aspects. Even if we only consider the report for the most advanced province in Pakistan, Punjab, it cited 11% absenteeism among children and 13% among teachers still in public schools. Only 31% of teachers had graduated from an institution, while 59% had obtained professional qualifications or bachelors degrees in education. Regarding school facilities, 79% of public schools had computer labs, and 83% had a library facility. Furthermore, only 2% of primary schools lacked toilets, while 4% were without drinking water. Other factors such as a lack of grants to schools, insufficient classrooms, fewer playgrounds, etc., are also detailed in the report [58]. Such surveys have been performed in the past with attention to specific institutions or regions and considering a limited set of institutional parameters [7,8]. In this research, a maximal set of influencing institutional parameters were included with a broader scope covering the regional level to measure overall, region-wide institutional performance. The results proved that the efficient provision of resources yields better educational results. It was also observed that the urban areas performed well compared to their rural counterparts due to the maximum availability of facilities and resources. Better school infrastructure and physical facilities increased student attendance, strengthened staff motivation, and improved student academic results.

There is always a link between school users (students, teachers) and school architecture. Past studies have demonstrated that a clean and safe learning environment plays a valuable role in academic achievement. Moreover, overcrowding of classrooms, toilets, laboratories, and dormitories, and dilapidated school structures create an uncomfortable school environment. Unhealthy school environments lower the morale of students, teachers, and parents, leading to higher dropout rates and poorer academic achievement [59–61].

Taking the 2030 agenda into consideration, formulating reliable education measures, measuring education disparities among districts, and investigating factors behind education disparities at the household level will all be imperative to the task of recommending effective policy options and the tackling the targets of the Sustainable Development Goals in earnest. This study will help support governance for performance monitoring, policy formulation, target-setting, evaluations, and reforms aimed at addressing the issues and challenges in education worldwide. Gaps in school participation can be better understood in terms of regional socio-economic, demographic, and geographic disparities. There were a few limitations to our study. Firstly, it only covered data for high schools in one province of Pakistan, and the results for other provinces may differ. Secondly, our model utilised a structured dataset, but the results may vary when unstructured or semi-structured data are utilised.

6. Conclusions

Whenever the government introduces educational policies that are based on analyses of performance not of a single school but of schools on a massive scale, region-wide—rather than individual–school performance measurements are a practical approach. The level of education in public institutions varies across all regions of Pakistan. The current disparities in access to education in Pakistan are mostly due to systemic regional differences and the distribution of resources. This study, therefore, sought to fill the gaps and emphasise the importance of region-wide measurements of school performance. A machine learning-based method was developed to generate results. It was revealed that aside from student academic results, other factors substantially impact the performance of any school institution. The present study focused on the importance of these other highly influential factors along with student academic results, e.g., teacher–student ratios, the number of schools per region, school locations in rural or urban areas, and the availability of classrooms, electricity in schools, furniture for students, open-air classes, computer labs, science labs, and school playgrounds. Our finding was that in Pakistan, discrepancies in the performance of educational institutions in different regions of the country are due to inequality in the distribution of resources, differences in essential facilities, the number of schools by region, and the influence of school location on motivation, literacy rates, and awareness levels in the local population. This study will help support governance for performance monitoring, policy formulation, target-setting, evaluations, and reforms to address the issues and challenges for education. Moreover, changing socio-economic factors may lead to different results. This research could be conducted on all schools—primary, middle, high—and even institutions of higher learning or in different regions of the nation. In the future, a few advanced ensemble-based machine learning algorithms such as extreme gradient boosting could be utilised in this domain.

Author Contributions: Conceptualization, T.M.A., M.M. and K.S.; methodology, T.M.A.,M.M., K.S., I.A.H.; software, I.A.H., M.U.S. and S.L.; validation, M.U.S. and S.L.; formal analysis, T.M.A., M.M. and K.S.; investigation, M.U.S. and S.L.; resources, M.U.S. and S.L.; data curation, T.M.A., M.M., K.S. and I.A.H.; writing—original draft preparation, T.M.A., M.M. and K.S.; writing—review and editing, T.M.A., M.M. and K.S.; visualization, M.U.S. and S.L.; supervision, M.U.S. and S.L. All authors have read and agreed to the published version of the manuscript.

Funding: This research received no external funding.

Institutional Review Board Statement: Not applicable.

Informed Consent Statement: Not applicable.

Data Availability Statement: The data associated with this article can be found in the online version at doi:10.17632/637d4s7vjh.1.

Conflicts of Interest: The authors declare no conflict of interest.

References

1. Tesema, M.T.; Braeken, J. Regional inequalities and gender differences in academic achievement as a function of educational opportunities: Evidence from Ethiopia. *Int. J. Educ. Dev.* **2018**, *60*, 51–59. [CrossRef]
2. Faisal, R.; Shinwari, L.; Mateen, H. Evaluation of the academic achievement of rural versus urban undergraduate medical students in pharmacology examinations. *Asian Pac. J. Reprod.* **2016**, *5*, 317–320. [CrossRef]
3. Jamil, M.; Mustafa, G.; Ilyas, M. Impact of school infrastructure and pedagogical materials on its academic performance: Evidence from Khyber Pakhtunkhwa. *FWU J. Soc. Sci.* **2018**, *12*, 42–55.
4. Ning, B.; Damme, J.V.; Noortgate, W.V.N.; Gielen, S.; Bellens, K.; Dupriez, V.; Dumay, X. Regional inequality in reading performance: An exploration in Belgium. *Sch. Eff. Sch. Improv.* **2016**, *27*, 642–668. [CrossRef]
5. Gbollie, C.; Keamu, H.P. Student academic performance: The role of motivation, strategies, and perceived factors hindering Liberian junior and senior high school students learning. *Educ. Res. Int.* **2017**, *2017*, 1–11. [CrossRef]
6. Honicke, T.; Broadbent, J. The influence of academic self-efficacy on academic performance: A systematic review. *Educ. Res. Rev.* **2016**, *17*, 63–84. [CrossRef]
7. Abdullah, N.A.; Bhatti, N. Failure in quality of academic performance of students in public sector schools of Sheikhupura. *J. Educ. Educ. Dev.* **2018**, *5*, 289–305. [CrossRef]
8. Fernandes, E.; Holanda, M.; Victorino, M.; Borges, V.; Carvalho, R.; Van Erven, G. Educational data mining: Predictive analysis of academic performance of public school students in the capital of Brazil. *J. Bus. Res.* **2019**, *94*, 335–343. [CrossRef]
9. Kassarnig, V.; Mones, E.; Bjerre-Nielsen, A.; Sapiezynski, P.; Dreyer Lassen, D.; Lehmann, S. Academic performance and behavioral patterns. *EPJ Data Sci.* **2018**, *7*, 1–16. [CrossRef]
10. Natek, S.; Zwilling, M. Student data mining solution–knowledge management system related to higher education institutions. *Expert Syst. Appl.* **2014**, *41*, 6400–6407. [CrossRef]
11. Gumus, S.; Chudgar, A. Factors affecting school participation in Turkey: An analysis of regional differences. *Compare* **2016**, *46*, 929–951. [CrossRef]
12. Chaudhry, R.; Tajwar, A.W. The Punjab Schools Reform Roadmap: A Medium-Term Evaluation. In *Implementing Deeper Learning and 21st Education Reforms*; Springer: Cham, Switzerland, 2021; pp. 109–128.
13. Aluko, R.O.; Daniel, E.I.; Oshodi, O.S.; Aigbavboa, C.O.; Abisuga, A.O. Towards reliable prediction of academic performance of architecture students using data mining techniques. *J. Eng. Des. Technol.* **2018**, *16*, 385–397. [CrossRef]
14. Nurliana, M.; Sudaryana, B. The influence of competence, learning methods, infrastructure facilities on graduate quality (case study (vocational high school) smkn 5 bandung indonesia). *Indones. J. Soc. Res.* **2020**, *2*, 18–43. [CrossRef]
15. Hameen, E.C.; Ken-Opurum, B.; Priyadarshini, S.; Lartigue, B.; Anath-Pisipati, S. Effects of school facilities' mechanical and plumbing characteristics and conditions on student attendance, academic performance and health. *Int. J. Civ. Environ. Eng.* **2020**, *14*, 193–201.
16. Belmonte, A.; Bove, V.; D'Inverno, G.; Modica, M. School infrastructure spending and educational outcomes: Evidence from the 2012 earthquake in Northern Italy. *Econ. Educ. Rev.* **2020**, *75*, 101951. [CrossRef]
17. Gul, M.; Shah, A.F. Assessment of physical school environment of public sector high schools in Pakistan and World Health Organization's Guidelines. *Glob. Reg. Rev.* **2019**, *4*, 238–249. [CrossRef]
18. Alasadi, S.A.; Bhaya, W.S. Review of data preprocessing techniques in data mining. *J. Eng. Appl. Sci.* **2017**, *12*, 4102–4107.
19. Kira, K.; Rendell, L.A. The feature selection problem: Traditional methods and a new algorithm. In Proceedings of the 10th National Conference on Artificial Intelligence, San Jose, CA, USA, 12–16 July 1992.
20. Robnik-Šikonja, M.; Kononenko, I. Theoretical and empirical analysis of ReliefF and RReliefF. *Mach. Learn.* **2003**, *53*, 23–69. [CrossRef]
21. Guyon, I.; Elisseeff, A. An introduction to variable and feature selection. *J. Mach. Learn. Res.* **2003**, *3*, 1157–1182.
22. Chandrashekar, G.; Sahin, F. A survey on feature selection methods. *Comput. Electr. Eng.* **2014**, *40*, 16–28. [CrossRef]
23. Xue, B.; Zhang, M.; Browne, W.N.; Yao, X. A survey on evolutionary computation approaches to feature selection. *IEEE Trans. Evol. Comput.* **2015**, *20*, 606–626. [CrossRef]
24. Tibshirani, R. Regression shrinkage and selection via the lasso. *J. R. Stat. Soc.* **1996**, *58*, 267–288. [CrossRef]
25. Khushi, M.; Shaukat, K.; Alam, T.M.; Hameed, I.A.; Uddin, S.; Luo, S.; Yang, X.; Reyes, M.C. A comparative performance analysis of data resampling methods on imbalance medical data. *IEEE Access* **2021**, *9*, 109960–109975. [CrossRef]
26. Alam, T.M.; Shaukat, K.; Mahboob, H.; Sarwar, M.U.; Iqbal, F.; Nasir, A.; Hameed, I.A.; Luo, S. A machine learning approach for identification of malignant mesothelioma etiological factors in an imbalanced dataset. *Comput. J.* **2021**, *00*, 1–12.
27. Alam, T.M.; Shaukat, K.; Hameed, I.A.; Khan, W.A.; Sarwar, M.U.; Iqbal, F.; Luo, S. A novel framework for prognostic factors identification of malignant mesothelioma through association rule mining. *Biomed. Signal Process. Control* **2021**, *68*, 102726. [CrossRef]
28. Shaukat, K.; Iqbal, F.; Alam, T.M.; Aujla, G.K.; Devnath, L.; Khan, A.G.; Iqbal, R.; Shahzadi, I.; Rubab, A. The impact of artificial intelligence and robotics on the future employment opportunities. *Trends Comput. Sci. Inf. Technol.* **2020**, *5*, 50–54.
29. Shaukat, K.; Alam, T.M.; Luo, S.; Shabbir, S.; Hameed, I.A.; Li, J.; Abbas, S.K.; Javed, U. A Review of Time-Series Anomaly Detection Techniques: A Step to Future Perspectives. In *Advances in Information and Communication, Proceedings of the Future of Information and Communication Conference (FICC 2021), Vancouver, BC, Canada, 29–30 April 2021*; Springer: Cham, Switzerland, 2021.

30. Shaukat, K.; Luo, S.; Abbas, N.; Mahboob Alam, T.; Ehtesham Tahir, M.; Hameed, I.A. An analysis of blessed Friday sale at a retail store using classification models. In Proceedings of the 4th International Conference on Software Engineering and Information Management (ICSIM 2021), Yokohama, Japan, 16–18 January 2021.

31. Shaukat, K.; Alam, T.M.; Ahmed, M.; Luo, S.; Hameed, I.A.; Iqbal, M.S.; Li, J.; Iqbal, M.A. A model to enhance governance issues through opinion extraction. In Proceedings of the 11th IEEE Annual Information Technology, Electronics and Mobile Communication Conference (IEMCON), Vancouver, BC, Canada, 4–7 November 2020.

32. Bashir, U.; Chachoo, M. Performance evaluation of j48 and bayes algorithms for intrusion detection system. *Int. J. Netw. Secur. Its Appl.* **2017**, *9*, 1–11. [CrossRef]

33. Srivastava, A.K.; Singh, D.; Pandey, A.S.; Maini, T. A novel feature selection and short-term price forecasting based on a decision tree (J48) model. *Energies* **2019**, *12*, 3665. [CrossRef]

34. Guenther, N.; Schonlau, M. Support vector machines. *Stata J.* **2016**, *16*, 917–937. [CrossRef]

35. Alam, T.M.; Iqbal, M.A.; Ali, Y.; Wahab, A.; Ijaz, S.; Baig, T.I.; Hussain, A.; Malik, M.A.; Raza, M.M.; Ibrar, S.; et al. A model for early prediction of diabetes. *Inform. Med. Unlocked* **2019**, *16*, 100204. [CrossRef]

36. Qi, Y. Random forest for bioinformatics. In *Ensemble Machine Learning*; Springer: Berlin/Heidelberg, Germany, 2012; pp. 307–323.

37. Boulesteix, A.L.; Janitza, S.; Kruppa, J.; König, I.R. Overview of random forest methodology and practical guidance with emphasis on computational biology and bioinformatics. *Wiley Interdiscip. Rev.* **2012**, *2*, 493–507. [CrossRef]

38. Niehaus, K.E.; Uhlig, H.H.; Clifton, D.A. Phenotypic characterisation of Crohn's disease severity. In Proceedings of the 37th Annual International Conference of the IEEE Engineering in Medicine and Biology Society (EMBC), Milan, Italy, 25–29 August 2015.

39. Louppe, G.; Wehenkel, L.; Sutera, A.; Geurts, P. Understanding variable importances in forests of randomized trees. *Adv. Neural Inf. Process. Syst.* **2013**, *26*, 431–439.

40. Dawer, G.; Barbu, A. Relevant ensemble of trees. *arXiv* **2017**, arXiv:1709.05545.

41. Rodriguez, J.J.; Kuncheva, L.I.; Alonso, C.J. Rotation forest: A new classifier ensemble method. *IEEE Trans. Pattern Anal. Mach. Intell.* **2006**, *28*, 1619–1630. [CrossRef]

42. Fegade, K.G.; Gupta, R.; Namdeo, V. Predictive model for multiclass classification of e-commerce data: An azure machine learning approach. *Int. J. Comput. Appl.* **2017**, *168*, 37–42.

43. Đurđević Babić, I. Machine learning methods in predicting the student academic motivation. *Croat. Oper. Res. Rev.* **2017**, *8*, 443–461. [CrossRef]

44. Borkin, D.; Némethová, A.; Michaľčonok, G.; Maiorov, K. Impact of Data Normalization on Classification Model Accuracy. *Res. Pap. Fac. Mater. Sci. Technol. Slovak Univ. Technol.* **2019**, *27*, 79–84. [CrossRef]

45. Alshdaifat, E.; Alshdaifat, D.; Alsarhan, A.; Hussein, F.; El-Salhi, S.M.D.F.S. The effect of preprocessing techniques, applied to numeric features, on classification algorithms' performance. *Data* **2021**, *6*, 11. [CrossRef]

46. Tsai, C.-F.; Chen, Y.C. The optimal combination of feature selection and data discretization: An empirical study. *Inf. Sci.* **2019**, *505*, 282–293. [CrossRef]

47. Lavangnananda, K.; Chattanachot, S. Study of discretization methods in classification. In Proceedings of the 9th International Conference on Knowledge and Smart Technology (KST), Pattaya, Thailand, 1–4 February 2017.

48. Alam, T.M.; Shaukat, K.; Hameed, I.A.; Luo, S.; Sarwar, M.U.; Shabbir, S.; Li, J.; Khushi, M. An investigation of credit card default prediction in the imbalanced datasets. *IEEE Access* **2020**, *8*, 201173–201198. [CrossRef]

49. Singh, D.; Singh, B. Investigating the impact of data normalization on classification performance. *Appl. Soft Comput.* **2020**, *97*, 105524. [CrossRef]

50. Weiss, S.; Xu, Z.Z.; Peddada, S.; Amir, A.; Bittinger, K.; Gonzalez, A.; Lozupone, C.; Zaneveld, J.R.; Vázquez-Baeza, Y.; Birmingham, A. Normalization and microbial differential abundance strategies depend upon data characteristics. *Microbiome* **2017**, *5*, 1–18. [CrossRef] [PubMed]

51. Alam, T.M.; Shaukat, K.; Mushtaq, M.; Ali, Y.; Khushi, M.; Luo, S.; Wahab, A. Corporate bankruptcy prediction: An approach towards better corporate world. *Comput. J.* **2020**, *65*, 1–16.

52. Ramírez-Gallego, S.; García, S.; Mouriño-Talín, H. Data discretization: Taxonomy and big data challenge. *Wiley Interdiscip. Rev.* **2016**, *6*, 5–21. [CrossRef]

53. Nguyen, H.T.; Phan, N.Y.K.; Luong, H.H.; Le, T.P.; Tran, N.C. Efficient discretization approaches for machine learning techniques to improve disease classification on gut microbiome composition data. *Adv. Sci. Technol. Eng. Syst.* **2020**, *5*, 547–556. [CrossRef]

54. Jishan, S.T.; Rashu, R.I.; Mahmood, A.; Billah, F.; Rahman, R.M. Application of optimum binning technique in data mining approaches to predict students' final grade in a course. In *Computational Intelligence in Information Systems*; Springer: Berlin/Heidelberg, Germany, 2015; pp. 159–170.

55. *The Global Gender Gap Report*; World Economic Forum: Cologne, Germany; Geneva, Switzerland, 2017.

56. Yasmeen, G.; Begum, R.; Mujtaba, B. Human development challenges and opportunities in Pakistan: Defying income inequality and poverty. *J. Bus. Stud. Q.* **2011**, *2*, 1.

57. Wang, Z.; Zhang, B.; Wang, B. Renewable energy consumption, economic growth and human development index in Pakistan: Evidence form simultaneous equation model. *J. Clean. Prod.* **2018**, *184*, 1081–1090. [CrossRef]

58. Shaukat, K.; Nawaz, I.; Aslam, S.; Zaheer, S.; Shaukat, U. *Student's Performance: A Data Mining Perspective*; LAP Lambert Academic Publishing: Saarbrücken, Germany, 2017.

59. Shaukat, K.; Nawaz, I.; Aslam, S.; Zaheer, S.; Shaukat, U. Student's performance in the context of data mining. In Proceedings of the 2016 19th International Multi-Topic Conference (INMIC), Islamabad, Pakistan, 5–6 December 2016; pp. 1–8.
60. Lian, B.; Kristiawan, M.; Fitriya, R. Giving creativity room to students through the friendly school's program. *Int. J. Sci. Technol. Res.* **2018**, *7*, 1–7.
61. Matshipi, M.; Mulaudzi, N.; Mashau, T. Causes of overcrowded classes in rural primary schools. *J. Soc. Sci.* **2017**, *51*, 109–114. [CrossRef]

Article

Online University Students' Perceptions on the Awareness of, Reasons for, and Solutions to Plagiarism in Higher Education: The Development of the AS&P Model to Combat Plagiarism

Muhammad Abid Malik [1], Ameema Mahroof [2] and Muhammad Azeem Ashraf [3,*]

[1]   Department of Education, Iqra University, Karachi 75500, Pakistan; m_abidmalik7@yahoo.com
[2]   Faculty of Education, University of the Punjab, Lahore 54590, Pakistan; ameema.mahroof@gmail.com
[3]   Research Institute of Education Science, Hunan University, Changsha 410082, China
*   Correspondence: azeem@hnu.edu.cn

Abstract: Academic plagiarism has remained a major concern for higher education institutions, as it hampers not only the quality of the teaching-learning process and research, but also the overall educational institution. This issue appears to be even more serious in online and distance education institutions. As a result, a qualitative study was conducted on an online university in Pakistan to investigate the determinants of academic plagiarism and to find ways to address this issue. The students were given an open-ended questionnaire to reflect their opinions on the awareness and understanding of plagiarism, its determinants, and ways to address it. The findings revealed that most of the 267 online university students had a poor awareness and understanding of plagiarism. Major reasons for students' plagiarism turned out to be a lack of a proactive approach to create awareness, an omission of citation conventions from course content, untrained teachers, a lack of strict penalties and their proper implementation, poor time management, a fear of failure, a lack of confidence, laziness, and a culture of plagiarism. The study proposes the Awareness, Support, and Prevention model (AS&P model) to address this issue in higher education institutions.

Keywords: plagiarism; ethics; academic dishonesty; online education; higher education; AS&P model; Pakistan

Citation: Malik, M.A.; Mahroof, A.; Ashraf, M.A. Online University Students' Perceptions on the Awareness of, Reasons for, and Solutions to Plagiarism in Higher Education: The Development of the AS&P Model to Combat Plagiarism. *Appl. Sci.* **2021**, *11*, 12055. https://doi.org/10.3390/app112412055

Academic Editor: João M. F. Rodrigues

Received: 13 October 2021
Accepted: 9 December 2021
Published: 17 December 2021

Publisher's Note: MDPI stays neutral with regard to jurisdictional claims in published maps and institutional affiliations.

1. Introduction

Although there is no universally accepted definition of plagiarism [1,2], as it is culturally and socially influenced [3], it may be taken as a type of fraud, where someone takes others' ideas or work and presents it as one's own [4]. The 7th edition of the APA (American Psychological Association) publication manual describes plagiarism as "the act of presenting the words, ideas, or images of another as your own" [5].

There are different types of plagiarism, and different ways in which a person may plagiarize [6]. Academic plagiarism is one of those types, which is done within academic work or assignments. Lathrop and Foss said that academic plagiarism occurs when one does not think or write by oneself, or does not give appropriate bibliographical references [7]. It is generally detected through similarity index software such as Turnitin or Urkund [8].

Academic plagiarism is considered unethical in academia, as it not only damages the overall quality of education, but also compromises merit. As a result, it often invites punitive and disciplinary actions from higher education institutions and regulatory bodies. Unfortunately, despite those disciplinary policies, preventive measures, and punitive actions, plagiarism continues to grow, especially in higher education [8,9]. One of the reasons for this may be the availability of modern tools and software that allow students to search relevant materials on the internet and other online sources with relative ease [10,11]. Increasing access to rephrasing software further adds to this issue. In this situation, educational institutions are forced to allocate a substantial amount of time, effort, and resources to address it [12]. Students' lack of awareness and understanding of plagiarism is another important reason. Bennett found that many students did not know the correct citation conventions that could lead them to unintentional plagiarism [13]. Fish and Hura echoed this, saying that the students who lacked clarity about plagiarism and its consequences were more likely to commit it [14].

The literature also mentions many other reasons for plagiarism. Some of them are a lack of awareness, language issues, a tight schedule and deadlines, teachers' attitudes, high competition and expectations, a fear of failure, and peers' influence [13,15–21]. Different types of pressures have also been pointed out as contributing factors [22–24]. Songsri-wittaya et al. [25] mentioned students' desire to obtain high marks, exam pressure, and stiff peer competition as the most important reasons. Time-related issues are also quite significant, as students often have to complete assignments and other tasks in a limited amount of time [2].

Another reason for plagiarism is the difficulty of tasks. Weak preventive and disciplinary actions and/or a lack of policy enforcement may also tempt the students to plagiarize. Harris [16] (p. 6) supported this by saying, "cheating in self-defence may appear rational in a highly competitive atmosphere, especially where students believe there are few operative punishments".

Plagiarism is also a serious issue in Pakistan [26–28]. Previous studies have revealed that most of the students are either unaware about it [29] or have a poor understanding [30]. Murtaza et al. [29] carried out a study investigating Pakistani university students' awareness and perceptions about plagiarism. They collected data from 25,742 students from 35 universities in Pakistan. The findings revealed that 94% of the participants were not aware of the plagiarism policy of the HEC (Higher Education Commission). Another study found that many students did not know about their own university's policies about it [30]. The education and training of the students about plagiarism and citation conventions are also generally overlooked. Fatima et al. found that students' lack of a proper education and training about plagiarism and an inadequate skillset were two of the most significant factors [8]. Apparently, the HEC and higher education institutions in Pakistan have failed in their responsibility to create an awareness and understanding of plagiarism and its policies. As a result, plagiarism continues to thrive in Pakistan.

Recent decades have seen a rapid growth of online and distance education across the world [31–33]. Although there have been quite a few studies about plagiarism amongst online and distance education students [34,35], such research about Pakistan is quite rare. Additionally, research about online university students is also quite limited. The current study tried to fill some of these research gaps by focusing on the students of the only online university in Pakistan. It had three main research objectives:

- To investigate the awareness and understanding of plagiarism among online university students;
- To find out the perceptions of online university students about its determinants;
- To provide suggestions on how to address the issue.

2. Materials and Methods

2.1. Research Method and Tool

The current study used a qualitative research approach to achieve the research objectives. The rationale for employing qualitative research is that it allows in-depth data gathering from the participants [36]. At the same time, there was a desire to collect data from a larger number of participants in a way where they could freely and independently express their thoughts; consequently, an open questionnaire was selected as the research tool. One can gather in-depth data from a large number of participants through an open questionnaire [36]. However, the drawback of this technique is that one cannot ask supplementary questions for further probing. Doró also used a similar technique (asking students to write a one-page opinion essay on plagiarism) for data collection [2].

In order to collect data, the students were sent three main questions through their official emails. Every question was further explained in parentheses for clarity. Those questions were, "What is plagiarism (understanding and awareness about the concept)?", "Why do you think the students plagiarize (reasons behind it)?", and "How can we address this issue (suggestions to control plagiarism)?". The email also included the aims and scope of the study and a consent letter for voluntary participation. The students were asked to send their responses in 10 days. Those who did not respond within that time were reminded again twice every 15 days.

There were no word limit requirements to write reflections, and the participants were encouraged to elaborate as much as possible. They were asked to send their responses in MS Word files so that the researchers could directly use them in NVivo.

To ensure research ethics, permission was granted from the university to conduct the study. All the participants were asked to sign a consent letter for their voluntary participation. They were clearly told about the purpose of the research and that their participation was voluntary. They were also assured of their anonymity and that their reflective writings would only be used for research purposes.

2.2. Population and Sample

Data were gathered from an online university in Pakistan. This university is the first completely online university in the country. The students belong to three different departments (management sciences, computer science, and education) in both undergraduate and graduate programs. They were contacted through their university emails. Data were collected from the students studying during the Fall 2019 semester (October 2019–March 2020). During that semester, there were 14,743 active students in the university receiving virtual education. An email including the aims and scope of the study and a consent form for voluntary participation was sent to the students. A similar email was sent to the students twice every 15 days. Following the guidelines provided by the literature [37–39], the data analysis phase started as soon as the sample size of this study was considered appropriate for data analysis. In total, 300 students replied to the email to show their interest in participating the study. However, 267 participants filled the questionnaire and submitted it the authors.

Among these 267 students, 159 were male, and 108 were female. As it is an online university with many students working full time, their age ranged widely (19–52) with a mean age of 26 years, and 71% of the participants were from urban areas, while 29% were from rural areas.

2.3. Data Analysis Technique

The data were analyzed through qualitative data analysis software Nvivo 11. Initially, all data were imported into Nvivo. The researchers read all the writings one by one to search for broader themes and categories. Later, Nvivo was used to create codes and themes. Data visualization was also used for the determinants. In total, Nvivo generated 22 themes for determinants. The researchers again looked into those themes and grouped them into five broader categories.

Verbatim quotations were also used frequently, as they help in conveying the feelings and emotions of the participants more clearly and forcefully [40,41]. Two of the researchers independently selected significant verbatim quotations. They were later discussed by all three researchers. More significant and relevant ones were retained and reported in the article.

Frequencies and percentages were also used to present the data more clearly.

3. Results and Discussion

This section contains data findings and discussion. Other than background information, data analysis was carried out on three different lines: the awareness and understanding of online university students about plagiarism, their perceptions about its determinants, and how it can be addressed. These findings were further interpreted and discussed in light of the literature.

3.1. Online University Students' Awareness and Understanding of Plagiarism

The first question that the students were asked to write about was focused on their awareness and understanding of plagiarism: what was plagiarism to them? In order to establish how relevant and appropriate students' awareness and understanding of plagiarism was, it was important to first take one definition as the standard. After reviewing the literature, the researchers decided to use the definition in the 7th edition APA publication manual [5] as a standard because of its clarity, comprehensiveness, relevance, and significance. It is also used widely in most of the universities in Pakistan and abroad. It has three important components: (i) copying or lifting text, (ii) using not only words but also ideas of other authors, and (iii) presenting work as your own (not citing the original work properly). When a student showed a limited idea or understanding of plagiarism, but not the whole idea (e.g., using only the term "copy and paste"), it was categorized as a poor/incomplete understanding.

The analysis of students' responses showed that out of 267 students, only 41 (15.36%) could define it properly. The rest either had a completely wrong idea about plagiarism ($n = 37$, 13.86%) or an incomplete idea ($n = 189$, 70.79%).

An overwhelming majority (41%) used the term "copy and paste" to describe plagiarism. Other terms used by the students were "copy" (26%), "cheating" (15%), "writing without citations" (11%), and others (7%). Among others, "stealing others' ideas", and "presenting others' work as your own" were prominent terms used by the students. One student wrote, " . . . *[If I] copy something and then paste it in my assignment, then it is plagiarism*". Another wrote, "*Plagiarism is copying many lines from [the] Internet*". Some of the students' understanding of plagiarism appears to be defined by how different software is used to work to check plagiarism. They mentioned lifting five consecutive words as plagiarism. These students did not seem to realize that plagiarism was not only copying and pasting the text, but also using others' ideas without giving proper credit to them. Some did not see that copying and pasting could be academically ethical if it is cited properly.

The literature also points out that most of the students do not have a proper awareness and understanding of plagiarism [42] and citing conventions [13,43]. This shows that, if students do not know the meaning of plagiarism, it is difficult for them to avoid it. There seems to be either an inability of the universities to increase awareness and understanding or an unwillingness of the students to learn about it.

3.2. Why Do Students Plagiarize?

The second question was related to the determinants of plagiarism. Based on the responses of 267 students, Nvivo generated 21 themes (Figure 1). These themes were reported 421 times by the participants.

The researchers investigated the themes and grouped them into five broader categories: a lack of awareness and a poor understanding of plagiarism, weak management of the education system and institutional issues, academic pressures and barriers, personal and psychological reasons, and plagiarism becoming a trend. The details of these categories and their themes are as follows.

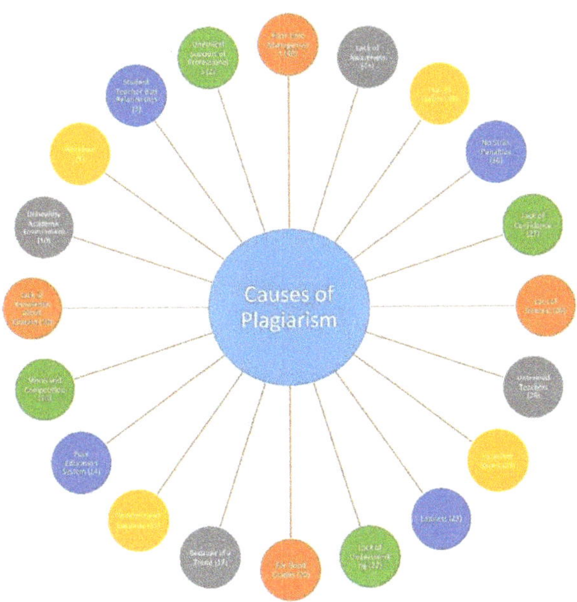

Figure 1. Determinants of plagiarism.

3.2.1. A Lack of Awareness and a Poor Understanding of Plagiarism

The most basic and fundamental cause seems to be the students' lack of awareness and understanding of plagiarism. Three Nvivo-generated themes were placed into this category: a lack of awareness, a poor understanding, and a lack of knowledge about citations.

The first theme was a lack of awareness, as 44 students wrote that they had a lack of awareness of plagiarism and the policies about it. Despite being university students, they said that they had not even heard about it. Their awareness about plagiarism, plagiarism policies, and the consequences of plagiarism was very limited. As one student wrote,

> " . . . .(In Pakistan) most students are unaware about plagiarism and its consequences. No one tells them about it. We do not find it anywhere. Some students [do not even] know the meaning of plagiarism. So most students are doing plagiarism without even knowing that they are doing plagiarism."

This situation is not surprising, as the literature also points this out [29,30]. Surprisingly, these are online university students who are more Internet savvy. The university has also provided information about plagiarism on both its website and the LMS (learning management system), but it seems that no proactive measures were taken to create an awareness of it.

The second theme in this category was a poor understanding of the concept. Twenty-two students said that the students did not have a proper understanding of the concept. This can also be connected with their responses to the first question, where the majority of students said that plagiarism was "copy", "copying", or "copy-paste" only. This confirms [42] (p. 643) the notion that students are "still apparently confused" about the concept of plagiarism.

The third theme was focused on students' knowledge about practices to avoid plagiarism, especially citation conventions. More than half of the students (149 out of 267) reported that they were never taught about plagiarism or citation conventions at any level. They stated that their lack of knowledge often led them to plagiarize content, even when they tried to avoid it. As one student wrote,

> *" … in citations and references … finding [the] right content and right method for citation[s] is [the] biggest problem. Many students do not know how to cite, how many words should be in inverted commas, [or] how to give reference[s]. Most of the time we think we [have] cited correctly, but in reality we [have] not, and we [have] committed] plagiarism."*

Another student explained how the lack of proper guidance about citations could play a role, saying,

> *"We [were] never taught about references and citations. There was no course in college, [or] even in university, [and] we do not have any course on this topic. Maybe [the institution and faculty] think that it is easy and we can learn by ourselves. But it is difficult. It is very difficult."*

Many studies have also pointed this out [13,43]. Often institutions think that citation conventions can be learned by the students themselves; however, in reality, citations and reference lists are quite difficult and confusing and should be taught properly.

These findings show the alarming situation of online university students' awareness and understanding of plagiarism in Pakistan. More than one fourth of them (76 out of 267) indicated their lack of awareness and a poor understanding of plagiarism and citing conventions, which has been a recurring theme in the literature [13,29,42]. Although the students had some idea about their own abilities and knowledge (or the lack of it) about plagiarism, it might be possible that many students are simply unaware about their lack of awareness. They might think that they know what plagiarism is, but their knowledge could be either completely wrong or incomplete.

3.2.2. Weak Management of the Education System and Institutional Issues

The second category that emerged from the data was the weak management of the education system and institutional issues, which either encouraged students to plagiarize or failed to prevent them from doing so. This category consists of six Nvivo-generated themes, i.e., a poor education system, an unhealthy academic environment, the unethical support of professionals, untrained teachers, a negative relationship between teachers and students, and a lack of strict penalties.

Fourteen students indicated a poor education system. Rather than pointing out any specific issue, they blamed the overall education system of the country. According to them, the blame goes to the overall educational system and culture in the country. These perceptions are similar to the authors in [44] who pointed out the academic and research culture in Pakistani higher education. One student expressed it in these words: *"[The] education system is poor. Problems are everywhere. Everything is bad"*. Another student echoed the same opinion, writing, *"It is difficult to blame one thing[;] everything in [Pakistani] education is bad"*.

In addition, six students put the blame on an unhealthy academic environment. One wrote, *" … students who try to follow rules [and be] honest in academics suffer"*. Similarly, talking about the academic environment in the university, another student wrote, *"It's [a] poor academic environment. We are not taught well, [as there are] just recorded lectures. No further explanation [required]"*. According to this student, as the students were not taught properly, they chose to plagiarize. Many students said that much greater efforts are needed in an online university to develop a strong academic culture that, to prevent such unethical practices, promotes not only academics but also moral values. Some researchers also indicated that online university students were more inclined to plagiarism [45].

Students also talked about the role of university personnel, especially faculty members, a negative relationship between teachers and students, and unethical support from professionals. Some students wrote that untrained teachers were key in the fast spread of plagiarism in Pakistan. The students mentioned that most of their teachers did not know about plagiarism, especially the proper meaning of plagiarism. One student expressed his personal experience as, *"My teacher said you should not write five consecutive words so most of us lifted the text and then rephrased every fourth or fifth word. I later found that it was still plagiarism".*

In addition, students mentioned "untrained and nonprofessional" research supervisors, who actually encouraged the students to commit plagiarism. One student wrote that his supervisor advised him to copy and paste the literature review and later use word rephrasing software to rephrase it. The students also reported unethical support from professionals and other office staff. A few students wrote that some teaching assistants and office staff would help the students in obtaining fake plagiarism reports.

Three students mentioned a negative teacher–student relationship. They pointed out that some teachers had a very aggressive and hostile attitude toward the students, and students did not feel comfortable asking them any questions. However, the standards and requirements of courses taught by those teachers were very strict, which compelled the students to plagiarize in order to produce "quality work". Howard also advocated that, rather than policies, educational institutions should focus on positive teacher–student relationships to eliminate plagiarism [46].

Another theme that stemmed from students' writings was a lack of strict penalties for plagiarism. Thirty-six students wrote about it and described how the students committing plagiarism often got away with either a simple warning or a lighter punishment. One student wrote, *"Penalties are flexible. Even though students are well aware of plagiarism, they still plagiarize and think they have [fewer] chances of being caught."* Another student further explained it in these words: *"Often professors just scold [students who plagiarize], [saying things] like 'shame on you' or 'do it again,' and that is it. Many times students [are] not caught".* One graduate student summed it up in these words, *"Soft penalties encourage students to plagiarize again and again".*

Researchers have been consistently vocal about the lack of penalties or light penalties for plagiarism as one of the key reasons for plagiarism [30]. In this situation, when students weigh in threats and benefits, they often find it worth taking a risk [16].

This clearly indicates that academic and nonacademic faculty members, education institutions, and even the overall education system in Pakistan is ill-equipped and ill-trained to address this issue. There also appears to be an issue with some of the teachers' attitudes, professionalism, and integrity. These factors along with the teachers' and universities' lenient approach toward plagiarism seem to have created an environment that is highly conducive to it.

3.2.3. Academic Pressures and Barriers

The most significant category was about different types of personal and academic pressures, and different barriers. These themes were mentioned by 140 students. These pressures and barriers appeared to force the students to plagiarize, as they could not manage or find a solution. This category contained six themes generated with Nvivo: poor time management, workload, stress from competition, a fear of failure, a desire for good grades, and language abilities.

Poor time management was the most frequently cited reason, as 49 students mentioned it. Time management substantially affects students across the world [47]. Many students have plagiarized to save time for other activities [2]. According to the participants, many students either wasted their time on other activities, or they simply could not manage their time properly. Therefore, they chose to plagiarize, as it would take less time to complete their work. One student wrote, *" . . . poor time management skills [will] in turn leave [students with] no choice [but] to cheat or plagiarize".*

Another participant described the students' attitude in these words,

"They do not manage their time very well and handle situations poorly. They focus on their [assignments] at [the] last moment [and work over a] very short time[;] it is very hard for students to do excellent work. So they try to get material online and finish their tasks."

Another determinant mentioned by five students was workload. Students said that their academic workload was too much to handle, so in order to manage heavy workload, they had to plagiarize. This has also been found in the literature [48]. Two students further indicated that a heavy workload was also related to time management. One student wrote that, even if they work hard, they cannot finish their workload on time. They had to cheat.

Sixty-three students mentioned stress from competition (10), a fear of failure (33), and a pressure to receive grades (20). This is understandable, as the education system in Pakistan measures students' performance in terms of marks and exam scores [49]. Those who receive higher marks tend to thrive in their professional life. This forces the students to aim for higher marks rather than cultivate their abilities. In order to compete with their fellow students, and to meet the required standards, some students decide to plagiarize. One student wrote, *"There is tough competition . . . we have to get high marks, [and that is it]."* Many students admitted that, in higher education, the fear of failure was one of the more important determinants for plagiarism. One of them wrote, *"If a student does not pass in an exam, people make fun of him. It is [shameful] for [a] student to get low grades. Many are afraid to fail. So they find other ways [such as] cheating to pass."* The literature also indicates the stress of competition, a fear of failure, and a pressure to receive good grades as more significant factors [13,15,18].

In addition to the themes mentioned above, 23 students mentioned the language barrier as one of the determinants. This seems to be a serious problem in countries such as Pakistan, where the language of the curriculum, instructions, and examinations is different from the daily language of communication [50]. This points to the fact that mostly books, assignments, and exams are in English. However, students communicate with each other in Urdu (the national language of Pakistan) or in a regional/ethnic language. As one student wrote, *"Students face difficulty in translating their thoughts from their mother tongue to English. So they use [a] 'copy-paste' method. It is not because we don't have ideas or knowledge [; rather,] it is because many cannot write their ideas in English."* English language writing is especially troublesome for nonnative speakers [50–52]. While conducting a study about Japanese students, Wheeler also found that a foreign-language-induced fear of failure was one of the main causes for plagiarism [53].

3.2.4. Personal and Psychological Reasons

Five Nvivo-generated themes were placed in the category of personal and psychological reasons. These themes were a lack of confidence, a lack of interest, laziness, looking for shortcuts and easy ways to finish their tasks, and the availability of the internet. These were identified by 27, 26, 23, 15, and 10 students, respectively.

A lack of self-belief and confidence is one of the more significant reasons for plagiarism [43]. Twenty-seven students echoed this, saying that, sometimes, the students were unsure of their own abilities and talent. They lacked confidence and self-belief, which would lead them to find alternative ways to finish their tasks. Sometimes, the lack of confidence is not due to academic problems, but to personal issues. One student wrote, *"Some students do not have confidence. They think others work better than them, [so] they chose to plagiarize."* Another student described it in these words, *"Some students are in a complex [situation, are afraid,] or have low confidence that they cannot write [a] good piece of text and feel hesitation, so they copy and paste in [their] assignments and other work."*

Twenty-six students identified the students' lack of interest as one of the key reasons. One student wrote, *"Often there is time. Students are active. But they don't have interest in the studies."* Another student said that many of those students who would plagiarize were quite active and smart. They had ample time to finish their tasks, but they would do other things. He wrote emphatically, *"They are more clever [and] more active. They find ways to cheat the teachers. It is not easy. They could do [a] better job in writing assignments, but they find no interest in writing [on] their own."*

Research about plagiarism shows that students often lacked interest in the given tasks. These students, due to their laziness, their disinterest, or other reasons, would look for shortcuts and easy ways to complete the task. Many students acknowledged that one of the causes of plagiarism was sheer laziness, as supported by the literature [52]. One student wrote, *"Often there is time, but students are lazy. They don't want to work."* Another student also reported that *"many students want to complete assignments [and receive] good marks, but they do not want to work. Plagiarism is an easy option for such students."*

Fifteen students said that many would plagiarize because they regard it as a quick and easy way of completing academic assignments and tasks. They did not specify if their desire to complete tasks in an easy way was due to laziness, a lack of interest, a lack of confidence, or an unavailability of modern tools. As a result, it was regarded as a separate theme.

The last theme in this category is the availability of the internet and other modern tools. Ten students cited the availability of the internet and other online tools and resources as one of the determinants in the spread of plagiarism. Multiple articles also indicate this [10,11], as modern tools and technologies help the students not only to find the relevant materials more easily but also to rephrase them to avoid detection. One student pointed this out, writing, *"[The] Internet has made writing easy. Students can find materials. Some software can rephrase it. It is not easy to catch them."* This leaves teachers and supervisors in a difficult situation. Until and unless these educators have read the original article and remember it clearly, it is difficult to catch this type of technologically savvy cheating.

3.2.5. Plagiarism Is Becoming a Trend

Seventeen students wrote that plagiarism had become a trend in Pakistan. They mentioned that, due to other determinants such as the lack of awareness and understanding, teachers' attitudes, and the institutional failure to pay attention to the relevant details, plagiarism was deeply rooted into much of the education system in Pakistan. This is very similar to what Horváth said about Hungary, that plagiarism was ingrained into the Hungarian education system [54]. It also appears to have become a routine for students. One of them wrote, *"No one [cares] about it. It is a trend. It is normal."*

This indicates that when systems fail to highlight an issue or to proactively try to eradicate it, with the passage of time, the issue may become part of people's lives and the general culture. Once that happens, people may no longer care about it. In other words, it becomes an acceptable evil. Worse still, it may become so common that some people may start bragging about it instead of feeling ashamed of their wrongdoing. One student said the same, *"It is trendy [;] many students do not feel shame. They tell others proudly."* This is a worrisome, deplorable situation that demands immediate measures and actions.

3.3. Addressing the Issue of Plagiarism

The last question was focused on how plagiarism in higher education settings in Pakistan can be prevented. The responses from the participants were coded into different categories, and themes were generated from the codes and categories. Three major themes that were considered important by the participants to combat plagiarism in higher education, i.e., awareness, support, and prevention, were established. Previous studies have also discussed these steps to control plagiarism [13,17,30,53,55]. This model integrates them with other details. Thus, the themes generated from the data from the current study and from previous literature with a similar scope guided the authors to develop the AS&P Model—the Awareness, Support, and Prevention Model (Figure 2).

The first dimension of the AS&P model is awareness, which represents the "what" part of the problem. It is about creating awareness about all the important "whats" relating to it (What is plagiarism? What are regulatory bodies and university policies and practices saying about it? What are its consequences? What are citation conventions?). The findings of the current study show that there is an acute lack of awareness about different aspects of plagiarism. The same has been noted in earlier studies [30]. Students strongly criticized the current approach of the universities, as one student wrote, *"Putting policies on [a] website and sitting [back without telling students about it] will not [create] awareness. [Universities] should reach [out] to the students [and] be active."*

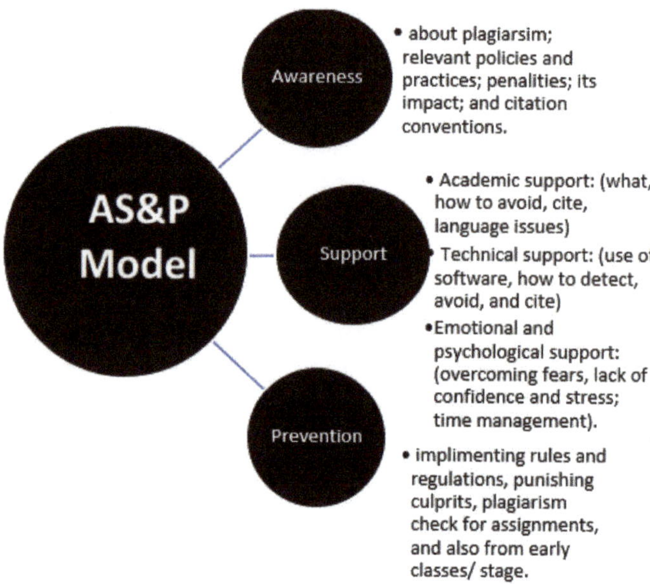

Figure 2. AS&P (Awareness, Support, and Prevention) Model.

As a lack of awareness has proven to be one of the most significant issues in this regard [13,42], many of the students recommended workshops, seminars, and training sessions to create an awareness and understanding of plagiarism. Interestingly, some of the students suggested that these workshops and training sessions should not be for the students only, but also for teachers. One student explained it, saying, *"When teachers know what [plagiarism is], they can guide students better."* They also suggested it be made a part of their course outlines.

The second dimension of the model is support, which represents the "how" part of the problem. Students wrote that supportive measures should be for both teachers and students. The different kinds of support suggested by the students have been divided into three categories: academic, technical, and emotional and psychological. In academic and technical support, the focus should be on the use of relevant software, how to detect and avoid plagiarism, and how to cite properly. Many of the students said that there should be support for language issues as well. One graduate student wrote at length about it, saying,

"[Universities] think [that the English of M.Phil students is good and that they] can write good English, but many cannot. We have [an] academic writing [program], but we don't do any practice; [it only consists of] lectures and information. It doesn't improve [one's] English. Universities in Pakistan should have special courses, [sessions] or workshops [to help students] in writing."

One student suggested that there should be a co-supervisor or member supervisory committee for language improvement. This person could help the students to improve their academic writing. Supervisors can focus on the technical and academic aspects, and the co-supervisor can focus on the language issues. In this way, the students could feel more comfortable with writing and avoid copying.

The students strongly suggested emotional and psychological support, as many of the determinants of plagiarism are related to emotional, psychological, and other nonacademic issues, such as a fear of failure, poor time management, laziness, and a lack of interest [2,15,52,53]. As a result, a positive attitude, counselling, and support from the universities, especially the teachers, are imperative to overcome these determinants. Leask very rightly pointed out that the prevention of plagiarism must not be seen as a war between academic staff and students, but as an opportunity for intercultural interactions and growth. He further said that preventing plagiarism therefore needs to be based on the principles of good teaching [55]. It was further suggested that rather than scolding the students for poor work, teachers should provide them with academic support and help in building confidence and gaining knowledge. Bashir and Malik also said that caring behavior exhibited by teachers in the classroom can help students not only with academic improvement, but also psychological, ethical, and moral improvements, which can lead to positivity and motivation amongst those students [56]. Some students also wrote about more active roles in student counseling centers in overcoming those psychological barriers. All these steps can greatly reduce the aforementioned emotional and psychological issues that have been some of the key factors of plagiarism.

The last dimension of the model is prevention. Despite creating awareness and providing supportive measures for students, some may still opt for foul play and cheating. Thus, the last dimension becomes pivotal. Universities must develop strict policies about plagiarism and enforce them in an active manner for both teachers and students. Students committing plagiarism and teachers involved in fostering plagiarism should be handled such that others will be deterred from that behavior. Most of the time, students plagiarize when they find that the benefits outweigh the risks [17]. If the risks (e.g., the punishments) are greater than the benefits, students are more likely to stay away from it.

Along with punitive actions, universities must also provide teachers with access to modern plagiarism-detection software to prevent plagiarism. Institutions should make software use mandatory for students' assignments, as well as for research work. Some students wrote that there was no culture of using plagiarism-detecting software for students' assignments or even research proposals. As one post-graduate student wrote,

> *"Even [for a] synopsis (research proposal), many students copy and paste, and they [pass]. Students don't worry about plagiarism at early stages, [in class assignments, course papers, or synopses] and then they are asked to come up with original work at thesis writing. It is not fair. Nobody cares about plagiarism and then suddenly they expect us to produce original work when thesis work starts. [Universities] should check [for] plagiarism all the time so that students learn about it, and are careful about it."*

The AS&P model appears to be quite comprehensive in eliminating or at least mitigating plagiarism in three different ways: creating awareness, providing support, and taking preventive measures. These measures are not only for students, but also for teachers. Plagiarism has become quite a menace in the current era, and it can only be controlled through a comprehensive approach.

4. Conclusions, Recommendations, and Implications

As indicated in the literature of other countries, the findings of the study confirm that plagiarism continues to remain a challenge at the higher education level in Pakistan. The findings of this study identify various determinants, such as an omission of citation conventions from course content, a lack of a proactive approach to create awareness, untrained teachers, a lack of strict penalties, poor time management, a fear of failure, a lack of confidence, laziness, and a culture of plagiarism. The students also questioned the role of universities and teachers, which ranged from reactive and laidback to promoting plagiarism. All of these factors appear to have created a culture of plagiarism in Pakistan.

Based on the suggestions given by the students, and consulting the literature on the same issue; this study provides a theoretical model, i.e., the AS&P model (Awareness, Support, and Prevention), to address this issue. There is a strong need to adopt a more proactive approach to create an awareness and understanding of plagiarism, the policies about it, and its consequences. Universities must first educate and train their teachers about it. Later, the students should be educated through seminars, workshops, and training sessions. Plagiarism and citing conventions should also be made a part of different course outlines. The students should be provided with support to combat different academic, technical, psychological, emotional, and other nonacademic issues, which often lead them to plagiarize. Teachers should adopt positive and caring behavior so that students can share their issues freely and seek solutions. Finally, both universities and teachers should adopt a zero-tolerance policy for it. Students and teachers involved in this practice should be punished so that others will be deterred. Plagiarism should also be checked for in students' assignments and research proposals/synopses so that they can learn to avoid it in early stages of the university career. It is worth mentioning that this model has been developed based on research findings and has not been tested in practice. Thus, researchers are encouraged to further investigate the experimental effectiveness of this model in reducing plagiarism in higher education settings.

Although this study is exclusively about Pakistan (more specifically about an online university in Pakistan), it has far-reaching implications. This model can be used in other countries, especially developing ones, with a similar context, a similar education system, and the same issues. Based on this model, universities can devise specific steps and strategies to reduce plagiarism and can emphasize different aspects of the model based on their context, issues, and needs.

5. Limitations and Further Research

The current study used an open questionnaire as a research tool. Although it helped in gathering qualitative data from a large number of participants, many statements made by the students required further explanation, and further probing through supplementary questions was not possible. Another qualitative study may be conducted using semi-structured or open interviews to probe this issue deeper.

Furthermore, it may be interesting to compare the causes of plagiarism between students studying online and students using a more conventional mode, as students in these two groups may hold different attitudes and mindsets in regard to plagiarism.

Finally, the study has devised the AS&P Model based on the literature and research findings. This model is theoretical and untested. It should be carefully implemented in an education system or universities to determine its effectiveness in reducing plagiarism. Therefore, future studies should experimentally investigate the effectiveness of this model in combating the issue of plagiarism in higher education.

Author Contributions: Conceptualization, M.A.M., A.M. and M.A.A.; methodology, M.A.M.; software, A.M.; validation, M.A.M., A.M. and M.A.A.; writing—original draft preparation, M.A.M.; writing—review and editing, A.M. and M.A.A. All authors have read and agreed to the published version of the manuscript.

Funding: The research received no external funding. The APC was funded by Research Funds for the Central Universities (Hunan University) in China, (Grant No. 531118010283).

Institutional Review Board Statement: This study was conducted according to the guidelines of the Declaration of Helsinki and approved by the Ethics Committee of Iqra University.

Informed Consent Statement: Written informed consent has been obtained from the participants.

Data Availability Statement: The data are available from the corresponding author at reasonable request.

Conflicts of Interest: The authors declare no conflict of interest.

References

1. Jameson, D.A. The ethics of plagiarism: How genre affects writers' use of source materials. *Bull. Assoc. Bus. Commun.* **1993**, *56*, 18–28. [CrossRef]
2. Doró, K. Why do students plagiarize? EFL undergraduates views on the reasons behind plagiarism. *Rom. J. Engl. Studies* **2014**, *11*, 255–263. [CrossRef]

3.   Hu, G.; Lei, J. Investigating Chinese University Students' Knowledge of and Attitudes toward Plagiarism from an Integrated Perspective. *Lang. Learn.* **2012**, *62*, 813–850. [CrossRef]

4.   Badke, W. Give Plagiarism the Weight It Deserves. *Online* **2007**, *31*, 58–60. Available online: https://www.researchgate.net/publication/292790712_Give_plagiarism_the_weight_it_deserves (accessed on 1 December 2021).

5.   American Psychological Association. *Publication Manual of the American Psychological Association*, 7th ed.; American Psychological Association: Washington, DC, USA, 2019.

6.   Ercegovac, Z.; Richardson, J.V. Academic dishonesty, plagiarism included, in the digital age: A literature review. *Coll. Res. Libr.* **2004**, *65*, 301–318. [CrossRef]

7.   Lathrop, A.; Foss, K. *Student Cheating and Plagiarism in the Internet Era: A Wake-Up Call*; Libraries Unlimited: Englewood, CO, USA, 2000.

8.   Fatima, A.; Sunguh, K.K.; Abbas, A.; Mannan, A.; Hosseini, S. Impact of pressure, self-efficacy, and self-competency on students' plagiarism in higher education. *Account. Res.* **2020**, *27*, 32–48. [CrossRef] [PubMed]

9.   Jereb, E.; Urh, M.; Jerebic, J.; Šprajc, P. Gender differences and the awareness of plagiarism in higher education. *Soc. Psychol. Educ.* **2017**, *21*, 409–426. [CrossRef]

10.   Comas-Forgas, R.; Sureda-Negre, J. Academic plagiarism: Explanatory factors from students' perspective. *J. Acad. Ethics* **2010**, *8*, 217–232. [CrossRef]

11.   Chang, C.M.; Chen, Y.L.; Huang, Y.; Chou, C. Why do they become potential cyber-plagiarizers? Exploring the alternative thinking of copy-and-paste youth in Taiwan. *Comput. Educ.* **2015**, *87*, 357–367. [CrossRef]

12.   Gullifer, J.; Tyson, G.A. Exploring university students' perceptions of plagiarism: A focus group study. *Stud. High. Educ.* **2010**, *35*, 463–481. [CrossRef]

13.   Bennett, R. Factors associated with student plagiarism in a post-1992 university. *Assess. Eval. High. Educ.* **2005**, *30*, 137–162. [CrossRef]

14.   Fish, R.; Hura, G. Students' perceptions of plagiarism. *J. Scholarsh. Teach. Learn.* **2013**, *13*, 33–45.

15.   Franklyn-Stokes, A.; Newstead, S.E. Undergraduate cheating: Who does what and why? *Stud. High. Educ.* **1995**, *20*, 159–172. [CrossRef]

16.   Harris, R. *The Plagiarism Handbook*; Pyrczak Publishing: Glendale, CA, USA, 2001.

17.   Park, C. In other (people's) words: Plagiarism by university students-literature and lessons. *Assess. Eval. High. Educ.* **2003**, *28*, 471–488. [CrossRef]

18.   Devlin, M.; Gray, K. In their own words: A qualitative study of the reasons Australian students plagiarize. *High. Educ. Res. Dev.* **2007**, *26*, 181–198. [CrossRef]

19.   Song-Turner, H. Plagiarism: Academic dishonesty or 'blind spot' of multicultural education? *Aust. Univ. Rev.* **2008**, *50*, 39–50.

20.   Rezanejad, A.; Rezaei, S. Academic dishonesty at universities: The case of plagiarism among Iranian language students. *J. Acad. Ethics* **2013**, *11*, 275–295. [CrossRef]

21.   Abbas, A.; Fatima, A.; Arrona-Palacios, A.; Haruna, H.; Hosseini, S. Research ethics dilemma in higher education. Impact of internet access, ethical controls, and teaching factors on student plagiarism. *Educ. Inf. Technol.* **2021**, *26*, 6109–6121. [CrossRef]

22.   Wilkinson, J. Staff and student perceptions of plagiarism and cheating. *Int. J. Teach. Learn. High. Educ.* **2009**, *20*, 98–105.

23.   Bacha, N.N.; Bahous, R. Student and teacher perceptions of plagiarism in academic writing. *Writ. Pedagog.* **2010**, *2*, 251–280. [CrossRef]

24.   Yazici, A.; Yazici, S.; Erdem, M.S. Faculty and student perceptions on college cheating: Evidence from Turkey. *Educ. Stud.* **2011**, *37*, 221–231. [CrossRef]

25.   Songsriwittaya, A.; Kongsuwan, S.; Jitgarum, K.; Kaewkuekool, S.; Koul, R. Engineering students' attitude towards plagiarism: A survey study. In Proceedings of the ICEE, ICEER Korea: International Conference on Engineering Education Research, Seoul, Korea, 12 September 2009.

26.   Sheikh, S. The Pakistan experience. *J. Acad. Ethics* **2008**, *6*, 283–287. [CrossRef]

27.   Shirazi, B.; Jafarey, A.M.; Moazam, F. Plagiarism and the medical fraternity: A study of knowledge and attitudes. *J. Pak. Med. Assoc.* **2010**, *60*, 269–273.

28.   Javaid, S.T.; Sultan, S.; Ehrich, J.F. Contrasting first and final year undergraduate students' plagiarism perceptions to investigate anti-plagiarism measures. *J. Appl. Res. High. Educ.* **2020**, *13*, 561–576. [CrossRef]

29.   Murtaza, G.; Zafar, S.; Bashir, I.; Hussain, I. Evaluation of student's perception and behavior towards plagiarism in Pakistani universities. *ActaBioethica* **2013**, *19*, 125–130. [CrossRef]

30.   Ramzan, M.; Munir, M.A.; Siddique, N.; Asif, M. Awareness about plagiarism amongst university students in Pakistan. *High. Educ.* **2012**, *64*, 73–84. [CrossRef]

31.   Goodman, J.; Melkers, J.; Pallais, A. *Can Online Delivery Increase Access to Education?* Working Paper 22754; National Bureau of Economic Research: Cambridge, MA, USA, 2016.

32.   Noreen, S.; Malik, M.A. Digital Technologies for Learning at Allama Iqbal Open University (AIOU): Investigating Needs and Challenges. *Open Prax.* **2020**, *12*, 39–49. [CrossRef]

33.   Malik, M.A.; Akkaya, B. Comparing the Academic Motivation of Conventional and Distance Education Students: A Study about a Turkish University. *Sir. Syed. J. Educ. Soc. Res.* **2021**, *4*, 341–351. [CrossRef]

34. Olt, M.R. Ethics and distance education: Strategies for minimizing academic dishonesty in online assessment. *Online J. Distance Learn. Adm.* **2002**, *5*, 1–7.
35. Ewing, H.; Anast, A.; Roehling, T. Addressing Plagiarism in Online Programmes at a Health Sciences University: A Case Study. *Assessment Eval. High. Educ.* **2016**, *41*, 575–585. [CrossRef]
36. Donald, A.; Jacobs, L.C.; Razavieh, A.; Sorensen, C.K. *Introduction to Research in Education*, 8th ed.; Hult Rinchart & Wiston: New York, NY, USA, 2010.
37. Krejcie, R.V.; Morgan, D.W. Determining sample size for research activities. *Educ. Psychol. Meas.* **1970**, *30*, 607–610. [CrossRef]
38. Patton, M.Q. *Qualitative Research and Evaluation Methods*; Sage: Thousand Oaks, CA, 2002.
39. Weller, S.C.; Vickers, B.; Bernard, H.R.; Blackburn, A.M.; Borgatti, S.; Gravlee, C.C.; Johnson, J.C. Open-ended interview questions and saturation. *PLoS ONE* **2018**, *13*, e0198606. [CrossRef]
40. Corden, A.; Sainsbury, R. Research Participants' Views on Use of Verbatim Quotations. Social Policy Research Unit. University of York; 2005. Available online: https://www.york.ac.uk/inst/spru/pubs/pdf/verb.pdf (accessed on 1 December 2021).
41. Corden, A.; Sainsbury, R. Exploring 'Quality': Research Participants' Perspectives on Verbatim Quotations. *Int. J. Soc. Res. Methodol.* **2006**, *9*, 97–110. [CrossRef]
42. Power, L.G. University students' perceptions of plagiarism. *J. High. Educ.* **2009**, *80*, 643–662. [CrossRef]
43. Dawson, J. Plagiarism: What's Really Going On? In *TL Forum 2004. Seeking Educational Excellence*; Murdoch University: Perth, Australia, 2004.
44. Hoodbhoy, P. Pakistan's Higher Education System-What Went Wrong and How to Fix It. *Pak. Dev. Rev.* **2009**, *48*, 581–594. [CrossRef]
45. Hart, L.; Morgan, L. Academic integrity in an online registered nurse to baccalaureate in nursing program. *J. Contin. Educ. Nurs.* **2010**, *41*, 498–505. [CrossRef]
46. Howard, R.M. Forget about Policing Plagiarism: Just Teach. The Chronicle of Higher Education, B24. Available online: https://abacus.bates.edu/cbb/events/docs/Howard_ForgeT.pdf (accessed on 16 November 2001).
47. Aduke, A.F. Time Management and Students Academic Performance in Higher Institutions, Nigeria-A Case Study of Ekiti State. *Int. Res. Edu.* **2015**, *3*, 1–12. [CrossRef]
48. Ehrich, J.; Howard, S.J.; Mu, C.; Bokosmaty, S. A Comparison of Chinese and Australian University Students' Attitudes Towards Plagiarism. *Stud. High. Educ.* **2016**, *41*, 231–246. [CrossRef]
49. Arum, R. Improve relationships to improve student performance. *Phi. Delta. Kappan.* **2011**, *93*, 8–13. [CrossRef]
50. Jamil, S.; Majoka, M.I.; Kamraan, U. Analyzing Common Errors in English Composition at Post-graduate Level in Khyber Pakhtunkhaw (Pakistan). *Bull. Educ. Res.* **2016**, *38*, 53–67.
51. Kharma, N. *A Contrastive Analysis of the Use of Verb Forms in English and Arabic*; Juliius Groos Verlog: Heidelberg, Germany, 1983.
52. Razera, D. Awareness, Attitude and Perception of Plagiarism among Students and Teachers at Stockholm University. Master's Thesis, Stockholm University, Stockholm, Sweden, 2011. Available online: https://www.diva-portal.org/smash/get/diva2:432681/FULLTEXT01.pdf (accessed on 1 December 2021).
53. Wheeler, G. Culture of minimal influence: A study of Japanese university students' attitudes toward plagiarism. *Int. J. Educ. Integr.* **2014**, *10*, 44–59. [CrossRef]
54. Horváth, J. A szakmai közösségnek tudnia kell ezekről a problémákról": A plágium kezelése a magyar egyetemi gyakorlatban" ["The academic public needs to know about these problems": Dealing with plagiarism in Hungarian higher education]. *Iskolakultúr* **2012**, *22*, 96–110.
55. Leask, B. Plagiarism, cultural diversity and metaphor-Implications for academic staff development. *Assess. Eval. High. Educ.* **2006**, *31*, 183–199. [CrossRef]
56. Bashir, S.; Malik, M.A. Caring Behavior of Teachers: Investigating the Perceptions of Secondary School Teachers and Students in Lahore. *Int. J. Innov. Teach. Learn.* **2020**, *6*, 63–78.

applied sciences

Brief Report

Identifying Engineering Undergraduates' Learning Style Profiles Using Machine Learning Techniques

Patricio Ramírez-Correa [1,*], Jorge Alfaro-Pérez [1] and Mauricio Gallardo [2]

[1] School of Engineering, Universidad Católica del Norte, Coquimbo 1781421, Chile; jalfaro@ucn.cl
[2] School of Business, Universidad Católica del Norte, Coquimbo 1781421, Chile; megallardo@ucn.cl
* Correspondence: patricio.ramirez@ucn.cl; Tel.: +56-9-77683692

Abstract: In a hybrid university learning environment, the rapid identification of students' learning styles seems to be essential to achieve complementarity between conventional face-to-face pedagogical strategies and the application of new strategies using virtual technologies. In this context, this research aims to generate a predictive model to detect undergraduates' learning style profiles quickly. The methodological design consists of applying a k-means clustering algorithm to identify the students' learning style profiles and a decision tree C4.5 algorithm to predict the student's membership to the previously identified groups. A cluster sample design was used with Chilean engineering students. The research result is a predictive model that, with few questions, detects students' profiles with an accuracy of 82.93%; this prediction enables a rapid adjustment of teaching methods in a hybrid learning environment.

Keywords: learning styles; machine learning; hybrid university teaching

Citation: Ramírez-Correa, P.; Alfaro-Pérez, J.; Gallardo, M. Identifying Engineering Undergraduates' Learning Style Profiles Using Machine Learning Techniques. *Appl. Sci.* **2021**, *11*, 10505. https://doi.org/10.3390/app 112210505

Academic Editor: João M. F. Rodrigues

Received: 23 September 2021
Accepted: 1 November 2021
Published: 9 November 2021

Publisher's Note: MDPI stays neutral with regard to jurisdictional claims in published maps and institutional affiliations.

1. Introduction

The COVID-19 pandemic has posed several challenges to higher education in teaching, learning, research collaboration and institutional governance [1]. Since it has been stated that the contexts in which learning occurs are crucial [2], one central research focus is examining and analysing learning styles in this new setting. This effort is even more critical given that the first reports about education in this new context indicate that the most frequent barrier is the difficulty in adjusting learning styles [3].

There are countless ways people process information from the environment, and individuals exhibit specific behaviours that allow them to learn efficiently [4]. They prefer interaction, assimilation, and information processing methods. This natural disposition or preference of the individual to learn and study is known as a learning style [5]. Since there are many learning styles among different individuals, it is a challenging task to determine and predict the learning style of an individual student. According to these ideas, adopting a standard pedagogy method is not appropriate to improve learning for all students. Therefore, it is essential to devise and adopt different pedagogies for different types of learners.

Universities in a pandemic face a hybrid teaching environment. The literature suggests that hybrid learning offers many benefits to students and faculties [6]. Nevertheless, this environment regularly changes scenarios and actors. Thus, the rapid identification of the learning styles of participants appears to be fundamental to achieving complementarity between traditional teaching and the application of new technologies. However, the extension of the existing learning style measurement instruments in the literature limits this rapid detection. In this context, is it possible to quickly detect the learning styles of university students of engineering? This question establishes the research problem for this study.

This study proposes developing a predictive model that identifies learning styles using analytical techniques. In the vein of a study that applied machine learning to reduce data capture times [7], our research is in line with works that reduce scales of learning styles [8–10]; however, we have advanced towards predicting a student profile associated with a learning style. The main practical contribution of this study to teaching in higher education is to allow administrators and educators to adopt different pedagogies for different groups of students quickly, improving students' academic results. Additionally, obtaining data on the learning styles of university students in the context of the pandemic in a developing country contributes to the generation of knowledge about this phenomenon for organisations and academics interested in higher education in the world.

In particular, the objective of this study is to generate a predictive model to detect undergraduates' learning style profiles quickly for making recommendations to teachers and managers to improve learning outcomes.

To achieve this objective, we proceed as follows. First, the following section describes the study background, including student learning profiles, student learning styles, and predictive analysis with decision trees. In the next section, the materials and methods are defined. Then, the results of the analyses are shown. Finally, the findings are discussed, limitations and future research lines are given, and conclusions are presented.

2. Background

2.1. Student Learning Profiles

Student learning profile refers to the preferred mode of learning as individuals. As a general consideration, the student obtains better results if tasks match with their skill and understanding (readiness), promote curiosity or passions (interest), and if the assignment fits their preferred learning profile [11]. Four overlapping categories of learning profile factors can be used to design a curriculum that fits students: group orientation, learning environment, cognitive style, and intelligence preferences [12]. Learning profiles have been studied from different perspectives and conditions. Specifically, student learning profiles have been studied in STEM education. At the K-12 level, [13] explored children's preference profiles on tangible and graphical robot programming, showing that student preference profiles are related to gender and age for both interfaces. At the undergraduate level, [14] examined the relationships between study-related burnout, learning profiles, study progressions, and study success. This research shows that learning profiles affect study-related burnout in higher education. Likewise, [15] studied academic performance prediction based on learning profiles in blended learning. Their results show that student learning profiles consisting of four online factors and three traditional factors have the highest predictive power of academic performance.

2.2. Student Learning Styles

Researchers agree that understanding student learning styles is a keystone for tailoring the teaching process, improving the satisfaction of educational needs, and enhancing learning experiences, especially in learning environments [16]. From a general point of view, since learning style is a component of the broader concept of personality [2], it may be related to specific personality traits. The Five-Factor Model has been widely used in the literature to measure personality traits. The FFM proposes five traits that capture the core domains of personality: conscientiousness, agreeableness, extraversion, neuroticism, and openness. Some studies have reviewed the relationship between psychological traits and the teaching–learning process. For example, [17] studied the predictive capacity of personality traits for teacher teaching styles in the Republic of China. The results indicate that personality traits contributed to the teaching styles of teachers beyond their gender, level of education, and perception of the quality of their students. On the other hand, the relationship between learning style and learners' personality was examined by [18]. This study reported that extroverted students tend to have an accommodative learning style. Finally, the concomitance between learning styles and psychological traits to explain learning to read English by Iranian students has been reported [19]. From a narrow point of view, the study of learning styles has generated considerable interest over the past three decades, leading to various models of learning styles based on how the learners adapt to multiple dimensions related to information reception and processing [5]. A review described 71 taxonomies of learning styles proposed in the literature [20].

The Felder–Silverman model is well-recognised in education [5,21]. In this model, learning styles refer to different strengths and preferences in acquiring and processing information [5]. We chose the Felder–Silverman model for this study due to two reasons. First, the model is widely accepted in engineering education [22]. Second, the measured scale of the model is reliable, valid, and suitable for engineering students [5,22].

The Felder–Silverman model classifies students by responding to four questions: What type of information does the learner preferentially perceive: sensory or intuitive? What type of sensory information is most effectively perceived: visual or verbal? How does the learner prefer to process information: actively or reflectively? How is the learner progressing in terms of sequential or overall comprehension? According to the answers, the learners are classified into four dimensions:

- D1—Perception: sensing (concrete thinker, practical, oriented towards facts and procedures) or intuitive (abstract thinker, innovative, oriented towards theories and underlying meanings).
- D2—Input: visual (learners prefer visual representations of material presented, such as pictures, diagrams, and flow charts) or verbal (learners prefer written and spoken explanations).
- D3—Processing: active (learners prefer to learn by trying things out, enjoy working in groups) or reflective (learners prefer to learn by thinking things through, such as working alone or with a single familiar partner).
- D4—Understanding: sequential (learners prefer to learn using a linear thinking process, learn in small incremental steps) or global (learners prefer to learn using a holistic thinking process, learn in giant leaps).

The Felder–Silverman model is operationalised by The Index of Learning Styles (ILS). The ILS is a 44-item questionnaire designed to evaluate preferences across the four dimensions of the Felder–Silverman model (active/reflective, sensing/intuitive, visual/verbal, and sequential/global). The Index of Learning Styles exposes an extensive application and formal validation [21]. For example, in management information systems, [23] studied the moderating impact of learning styles on the success of learning management systems using the Felder–Silverman model. Their results show that it is possible to improve the model performance through context-dependent moderators. Likewise, [24] studied preferred learning styles in an extensive undergraduate anatomy course (2,300 students). Their results suggest that anatomy students possess the predominant learning style dimensions seen in other STEM curricula.

2.3. Predictive Analysis with Decision Trees

A decision tree identifies a model that best fits the relationship between the attribute set and the class label of the input data. Specifically, decision trees have a hierarchical structure composed of a group of internal nodes and leaf nodes that classify a set of data by categorising them from the root node to some leaf node. Each internal node in the tree specifies a test condition that evaluates one or more attributes. Each descendant branch of the tree represents a sequence of decisions made by the model to determine the class membership of a new unclassified entity.

Unlike other classification models considered black boxes, decision trees are white-box models that allows someone to see why the model classifies in one way or another or to argue such a classification.

Different techniques have been developed to induce decision trees in the machine learning community. One of the pioneering works came from Quinlan with the ID3 algorithm [25]. This algorithm generates decision trees based on the information obtained from training examples and then uses them to classify the test set. The dataset generally has nominal attributes to perform the classification task with non-missing values.

The C4.5 algorithm is an extension of the ID3 algorithm Quinlan introduced to improve certain deficiencies [26]. Since it was not intended for numerical attributes and did not use pruning to reduce overtraining, the C4.5 algorithm uses a new calculation that allows the measuring of a gain ratio. It also handles attributes with continuous values. Finally, C4.5 employs a pruning technique to reduce the error rate. This technique reduces the size of the tree by removing sections that may be based on erroneous or missing data, thus reducing the complexity of the tree and improving its classification power.

There are several advantages in using decision trees [27]. First, the graphical representation of decision trees is intuitive when there are a reasonable number of nodes for users unfamiliar with the subject. In general, this favours transparency and decision making between professionals from different areas. Second, decision trees, unlike other techniques, are helpful for regression and classification problems. Third, the algorithms for creating decision trees are very flexible with the data. They can handle nominal, ordinal, and numeric data. Additionally, many of these algorithms can take missing and even errored values, which is useful for saving time in the data-cleaning process.

On the other hand, using decision trees also has its disadvantages. First, the tree can become complex when the data include nominal variables with many categories or several numerical variables. As a result, it tends to overfit the data with which it was trained. However, techniques such as pruning and setting growth limits solve this problem. Second, they are sensitive to irrelevant characteristics and variability in the data. Slight variations in the data can result in a completely different tree. Cross-validation procedures are used to avoid this problem. Finally, the process of building a decision tree can take a significant amount of time. This issue usually happens when there are many characteristics of each observation due to the algorithms in each iteration that compare which best divides the data.

3. Materials and Methods

3.1. Scales and Attributes

The measurement scales of this study have been extensively tested in previous research [5]. The ILS was used to measure learning styles. Additionally, we used the FFM to measure the students' personalities. The FFM was implemented through the Spanish translation of the Ten-Item Personality Inventory [28]. In particular, [28] validated the Ten-Item Personality Inventory in the Spanish language in a sample of 1,181 Spanish adults. Overall, [28] reported that the scale exhibited acceptable psychometric properties for measuring the FFM in terms of reliability, agreement, factor structure, and convergence with the traditional scale.

Finally, based on the previous literature [29–31], we develop a list of the possible attributes related to the learning style: gender, age, learning style in which the student is proficient, academic performance, use of social networking sites, and previous technical education. The learning style in which the student perceives himself or herself to be proficient was directly consulted through a single question based on a previous study [29].

3.2. Data

For the empirical study, a cluster sampling design was used to gather the data of Chilean engineering students. Two control variables were employed to define the cluster sampling: academic programmes (industrial engineering, computer engineering, and information technology engineering) and the level of the courses (five levels). We selected these two cluster variables because they are the two relevant institutional characteristics that reflect the distribution of students in the target population.

The data were obtained through an online questionnaire for students belonging to an engineering school in Coquimbo (Chile). The survey was conducted in August 2021. All the participants gave their notified consent before they contributed to the study. The research was conducted following the Declaration of Helsinki. The protocol was approved by the Ethics Committee of the Universidad Católica del Norte (Resolution No. 21 of 22 June 2021), guaranteeing the safeguard of the ethical principles for research declared by the committee. A total of 268 surveys were completed for the study. Most of the completed surveys were finished by males (74 per cent), and the average age was 20.9 years old. Regarding the academic background of the study's participants, 34.8 per cent were from the computer engineering major (93 students), 49.6 per cent from the industrial engineering major (133 students), and the remaining percentage from the information technology engineering major (42 students). Seventy-two students were from year 1, 69 from year 2, 39 from year 3, 51 from year 4, and 37 from year 5. The median of these students' grades was between 5 to 5.5, on a scale of 1 to 7 being seven the maximum. See Table 1 for other details regarding the distribution of the sample according to some variables of interest. The scatter diagram in Figure 1 shows the relationship between gender-separated personality traits.

Table 1. Distribution of the attributes of interest.

Attribute		N	%
Gender			
	Male	198	74
	Female	70	26
Learning style that the student was skilled at			
	Reading	11	4
	Writing	43	16
	Listening	26	10
	Doing	188	70
	Total	268	100
Age		Mean 20.9 ± 2.6	
		Range 18–46 years	

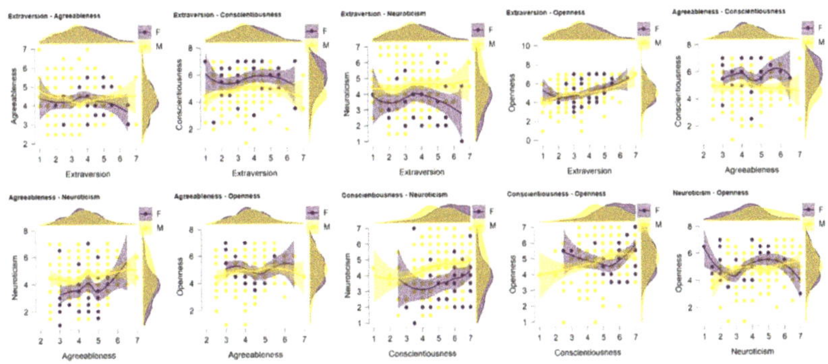

Figure 1. Scatter plot between gender-separated personality traits.

4. Results

4.1. Cluster Analysis

Following the proposal of [32], we used a k-means clustering algorithm to categorise students based on their style learning preferences. Table 2 and Figure 2 show the cluster analysis results. This method identified two clusters when optimised concerning the silhouette value. We used the average silhouette method to determine the number of clusters. This approach assesses the quality of clustering by determining the extent to which each object resides in its cluster. An elevated average silhouette width indicates a good clustering. This method calculates the average silhouette of the observations at different values of k. The optimum number of k clusters maximises the average silhouette over a range of possible values for k [33]. Figure 3 shows the clusters concerning the dimension of students' learning style preferences graphically.

Table 2. Results of k-means clustering analysis.

	Cluster 1 (56.4%)			Cluster 2 (43.6%)		
	Z-Score	Mean	SD	Z-Score	Mean	SD
D1: Perception	0.138	−1.694	3.093	−0.471	−5.167	2.804
D2: Input	0.547	−0.919	3.043	−0.487	−4.153	2.188
D3: Processing	0.565	0.097	3.096	−0.474	−3.250	2.546
D4: Understanding	0.550	0.145	2.876	−0.431	−2.597	2.360

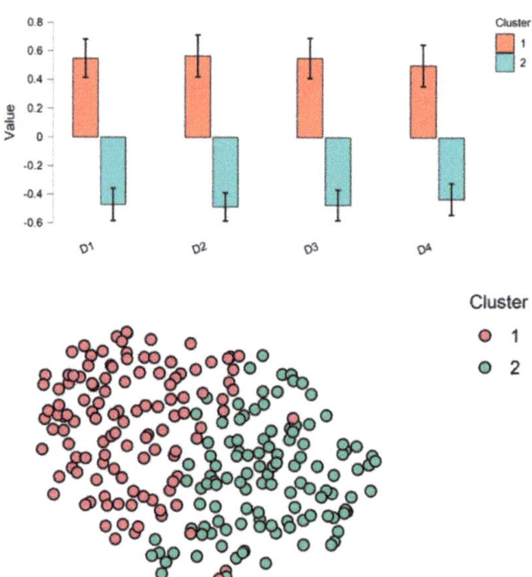

Figure 2. Cluster mean plots and t-SNE cluster plot.

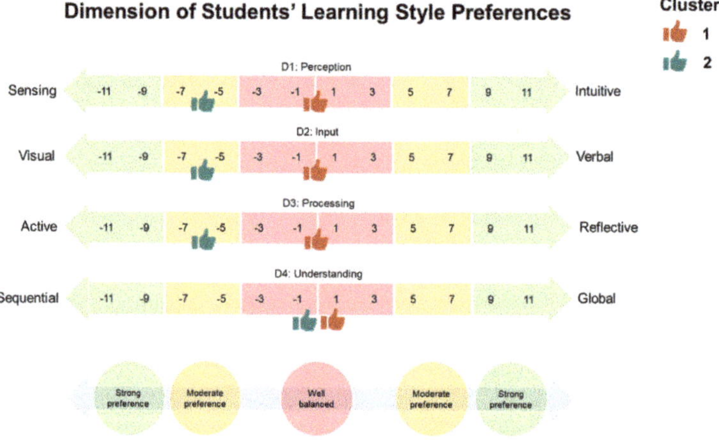

Figure 3. Clusters and learning styles.

Cluster 1 is the largest and corresponds to 56.4% of the sample. In this cluster, regarding learning style, students have higher Z-scores in all the dimensions and a well-balanced preference in the dimensions of perception, input, processing, and understanding between sensing–intuitive, visual–verbal, active–reflective, and sequential–global poles, respectively. The dimension with the lowest mean in this cluster is perception.

Cluster 2 is the smallest in size and corresponds to 43.6% of the sample. In this cluster, regarding learning style, students have a moderate preference in the dimensions of perception, input, and processing to sensing, visual, and active poles, respectively. Additionally, students have a well-balanced preference in the dimension of understanding between sequential and global poles. Similar to Cluster 1, the dimension with the lowest mean in this cluster is perception.

In qualitative terms, these results indicate that Cluster 1 is characterised by being more intuitive, verbal, reflective, and global in the learning profile. Cluster 2, in contrast, is typified by being more sensitive, active, and sequential in the learning profile.

4.2. Predictive Analysis

The prediction of the cluster associated with preferences of learning was conducted using decision trees. Specifically, we employed the C4.5 algorithm in this study [26], making decision trees from training data collection using information criteria. In addition, we used a grid optimisation strategy to set the parameters. This procedure indicated accuracy as division criteria and a maximum depth of nine.

Furthermore, to avoid overfitting, the analysis was performed using a 10-fold cross-validation with a training sample of 85%; the remaining sample of 15% was reserved to test the model with unseen data.

Lastly, the two-class criteria measured the performance prediction: sensitivity $= TP/(TP + FN)$, specificity $= TN/(TN + FP)$, precision $= TP/(TP + FP)$, and accuracy $= (TP + TN)/(TP + FP + FN + TN)$; where TP is a true positive, TN is a true negative, FP is a false positive, and FN is a false negative.

The prediction outcomes in Table 3 reveal that the method performs well regarding selecting the cases that need to be chosen, with an accuracy of $72.20 \pm 8.82\%$. In addition, the prediction outcomes in Table 4 reveal that the method performs well regarding selecting the cases of unseen data that need to be chosen, with an accuracy of 82.93%. Figure 4a–c shows the decision tree model.

Table 3. Confusion matrix for prediction performance of training data.

Accuracy: 72.20%	Cluster 1 (True)	Cluster 2 (True)	Class Precision
Cluster 1 (pred.)	89	35	71.77%
Cluster 2 (pred.)	33	87	72.50%
Class recall	72.95%	71.31%	

Table 4. Confusion matrix for prediction performance of unseen data.

Accuracy: 82.93%	Cluster 1 (True)	Cluster 2 (True)	Class Precision
Cluster 1 (pred.)	15	3	83.33%
Cluster 2 (pred.)	4	19	82.61%
Class recall	78.95%	86.36%	

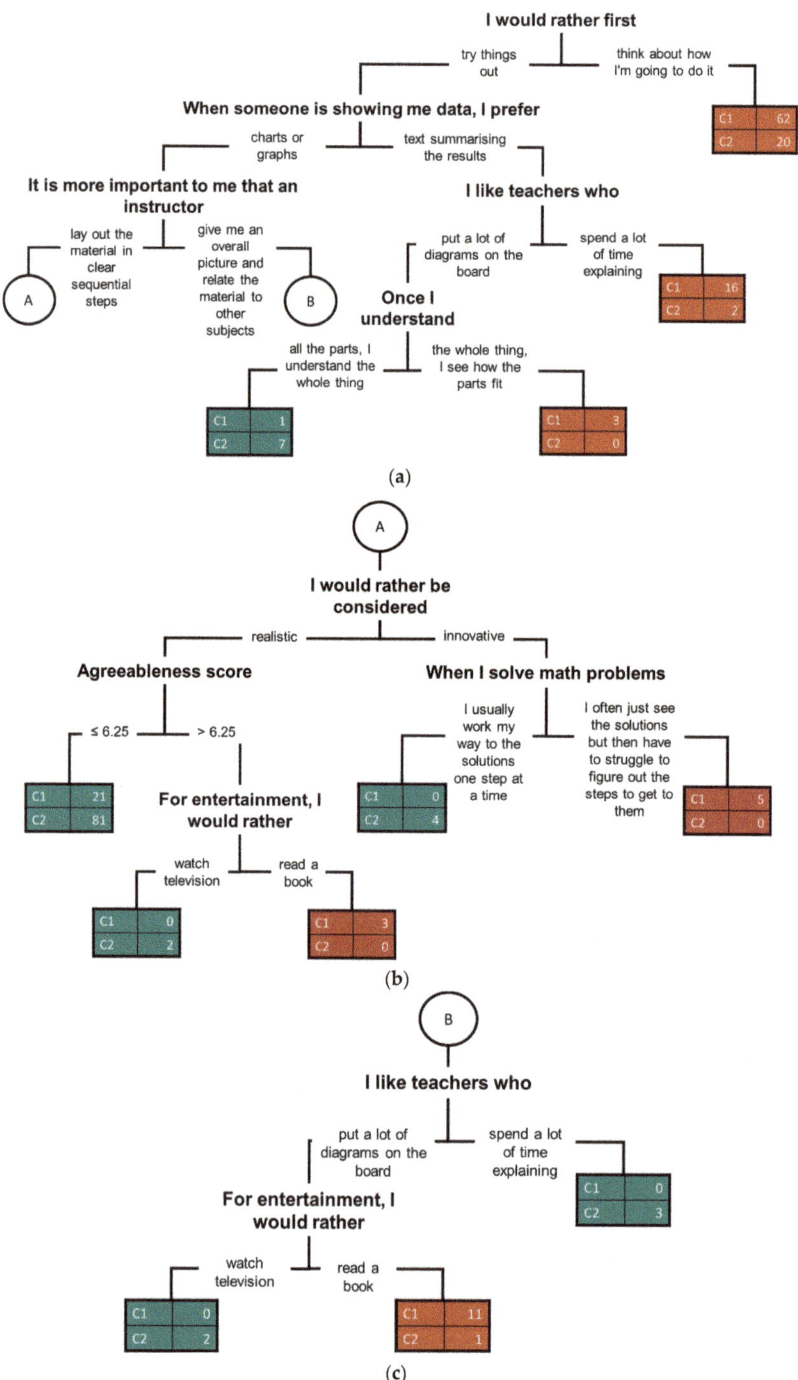

Figure 4. (a) Decision tree model; (b) decision tree model (part A); (c) decision tree model (part B).

5. Discussion

This paper generated a model to detect undergraduates' profiles based on swift learning styles. A machine learning process founded on a C4.5 algorithm generates a predictive model to classify students into two profiles. The two student learning style profiles were previously obtained from k-means clustering analysis. The result of two distinct learning style profiles is a remarkable finding: both tend towards the sensing, visual, and active poles. This fact is consistent with the literature of both engineering students [34] and students of other disciplines [35,36]. In a sample of Chilean engineering students, [23] found that most of the students were oriented to the sensing (84 per cent), visual (76 per cent), and active (70 per cent) poles. In a sample from a university in Mexico, [37] reported that engineering students tended towards the sensing (82 per cent), visual (90 per cent), and active (67 per cent) poles. As well, in a sample of manufacturing engineering students in Ireland, [38] found a bias towards the sensing (78 per cent), visual (per cent), and active (70 per cent) poles. In undergraduate business students, [35] reported that these learners tended toward the sensory (70 per cent), visual (68 per cent), and active (64 per cent) poles. Finally, a study by [39] discovered among industrial engineering students in Brazil the trend towards sensing (70 per cent), visual (73 per cent), and active (66 per cent) poles. Although there is the recent proposal of [32], we do not find student cluster reports associated with their learning styles based on the ILS, hence the importance of these results.

Although our study is in the line of works to reduce the scale of learning styles [8–10], these research findings distinguish an engineering student profile using a few questions of the ILS. Furthermore, the results suggest that attributes such as gender, age, academic performance, previous education, behaviour in social networks, perception of preferences to learn, or psychological traits are not appropriate to predict the student's profile. The exception is agreeableness; this trait discriminates between the clusters.

In particular, the results indicate that the model's predictive capacity is good (82.93%). Additionally, we highlight the rapidity of application due to this being essential in a hybrid environment: the model classifies 65% of students with no more than five questions, respectively. This decision tree model is the basis for the design of a computerised survey system for rapid discrimination.

Two limitations of this study should be stated. First, the analysis was conducted on a relatively small sample in one unit, which does not directly extrapolate the results. Second, a limited set of student attributes was used to predict their learning style profile, and, as a result, other attributes may predict this profile with greater accuracy.

Future studies should go three ways. First, to extend the analysis to a more significant sample. Second, to explore in other samples of engineering students in emerging economies the two profiles discovered in this investigation. Third, using new student attributes to predict their learning style profile, including personal values.

6. Conclusions

This study proposes an alternative to rapidly detect the learning styles of university students in a changing environment, such as hybrid teaching in the pandemic. The results indicate that based on a decision tree model, it is possible to determine, in a couple of questions and with acceptable performance, the profile of the students in a hybrid teaching activity. Moreover, this prediction enables a quick adjustment of teaching methods in a new environment.

Author Contributions: Conceptualisation, P.R.-C. and J.A.-P.; methodology, M.G.; software, P.R.-C.; validation, P.R.-C. and J.A.-P.; formal analysis, P.R.-C.; investigation, P.R.-C., M.G. and J.A.-P.; resources, P.R.-C.; data curation, P.R.-C.; writing—original draft preparation, P.R.-C.; writing—review and editing, P.R.-C. and M.G.; visualisation, P.R.-C. and J.A.-P.; supervision, P.R.-C.; project administration, P.R.-C.; funding acquisition, P.R.-C. All authors have read and agreed to the published version of the manuscript.

Funding: This research was funded by UNIVERSIDAD CATÓLICA DEL NORTE, grant FID 2021 DGPRE. 041/2021 and ANID, grant FONDECYT 1210130. The APC was funded by UNIVERSIDAD CATÓLICA DEL NORTE.

Institutional Review Board Statement: The study was conducted according to the guidelines of the Declaration of Helsinki and approved by the Ethics Committee of UNIVERSIDAD CATÓLICA DEL NORTE (Resolution No. 21 of 22 June 2021).

Informed Consent Statement: Informed consent was obtained from all subjects involved in the study.

Data Availability Statement: The data presented in this study are available in https://doi.org/10.6084/m9.figshare.16571807.v1.

References

1. Mok, K.H.; Xiong, W.; Ke, G.; Cheung, J.O.W. Impact of COVID-19 pandemic on international higher education and student mobility: Student perspectives from mainland China and Hong Kong. *Int. J. Educ. Res.* **2021**, *105*, 101718. [CrossRef]
2. Hawk, T.F.; Shah, A.J. Using learning style instruments to enhance student learning. *Decis. Sci. J. Innov. Educ.* **2007**, *5*, 1–19. [CrossRef]
3. Baticulon, R.E.; Sy, J.J.; Alberto, N.R.I.; Baron, M.B.C.; Mabulay, R.E.C.; Rizada, L.G.T.; Tiu, C.J.S.; Clarion, C.A.; Reyes, J.C.B. Barriers to Online Learning in the Time of COVID-19: A National Survey of Medical Students in the Philippines. *Med. Sci. Educ.* **2021**, *31*, 615–626. [CrossRef] [PubMed]
4. Willingham, D.T.; Hughes, E.M.; Dobolyi, D.G. The Scientific Status of Learning Styles Theories. *Teach. Psychol.* **2015**, *42*, 266–271. [CrossRef]
5. Felder, R.; Silverman, L. Learning and teaching styles in engineering education. *Eng. Educ.* **1988**, *78*, 674–681. [CrossRef]
6. Ora, A.; Sahatcija, R.; Ferhataj, A. Learning Styles and the Hybrid Learning: An Empirical Study about the Impact of Learning Styles on the Perception of the Hybrid Learning. *Mediterr. J. Soc. Sci.* **2018**, *9*, 137–148. [CrossRef]
7. Salazar-Concha, C.; Ramírez-Correa, P. Predicting the Intention to Donate Blood among Blood Donors Using a Decision Tree Algorithm. *Symmetry* **2021**, *13*, 1460. [CrossRef]
8. Yagci, M. Development of an Artificial Intelligence Based Computerized Adaptive Scale and Applicability Test. In Proceedings of the International Conference on Artificial Intelligence and Applied Mathematics in Engineering, Antalya, Turkey, 18–20 April 2020; Springer: Cham, Switzerland, 2020; pp. 292–303.
9. Mwamikazi, E.; Fournier-Viger, P.; Moghrabi, C.; Barhoumi, A.; Baudouin, R. An adaptive questionnaire for automatic identification of learning styles. In Proceedings of the International Conference on Industrial, Engineering and Other Applications of Applied Intelligent Systems, Kaohsiung, Taiwan, 3–6 June 2014; Springer: Cham, Switzerland, 2014; pp. 399–409.
10. Jing, Y.P.; Li, B.; Chen, N.; Li, X.F.; Hu, J.; Zhu, F. The discrimination of learning styles by Bayes-based statistics: An extended study on ILS system. *Control Intell. Syst.* **2015**, *43*, 68–75. [CrossRef]
11. Tomlinson, C.A.; Brighton, C.; Hertberg, H.; Callahan, C.M.; Moon, T.R.; Brimijoin, K.; Conover, L.A.; Reynolds, T. Differentiating instruction in response to student readiness, interest, and learning profile in academically diverse classrooms: A review of literature. *J. Educ. Gift.* **2003**, *27*, 119–145. [CrossRef]
12. Tomlinson, C.A. *How TO Differentiate Instruction in Mixed-Ability Classrooms*; ASCD: Alexandria, VA, USA, 2001; ISBN 0871209179.
13. Sapounidis, T.; Stamovlasis, D.; Demetriadis, S. Latent class modeling of children's preference profiles on tangible and graphical robot programming. *IEEE Trans. Educ.* **2018**, *62*, 127–133. [CrossRef]
14. Asikainen, H.; Salmela-Aro, K.; Parpala, A.; Katajavuori, N. Learning profiles and their relation to study-related burnout and academic achievement among university students. *Learn. Individ. Differ.* **2020**, *78*, 101781. [CrossRef]
15. Lu, O.H.T.; Huang, A.Y.Q.; Huang, J.C.H.; Lin, A.J.Q.; Ogata, H.; Yang, S.J.H. Applying learning analytics for the early prediction of Students' academic performance in blended learning. *J. Educ. Technol. Soc.* **2018**, *21*, 220–232.
16. Cheng, Y.M. Extending the expectation-confirmation model with quality and flow to explore nurses' continued blended e-learning intention. *Inf. Technol. People* **2014**, *27*, 230–258. [CrossRef]
17. Zhang, L. Do personality traits make a difference in teaching styles among Chinese high school teachers? *Pers. Individ. Dif.* **2007**, *43*, 669–679. [CrossRef]
18. Li, M.; Armstrong, S.J. The relationship between Kolb's experiential learning styles and Big Five personality traits in international managers. *Pers. Individ. Dif.* **2015**, *86*, 422–426. [CrossRef]
19. Foroozandehfar, L.; Khalili, G.F. On the relationship between Iranian EFL learners' reading fluency, their personality types and learning styles. *Cogent Arts Humanit.* **2019**, *6*, 1681347. [CrossRef]
20. Coffield, F.; Moseley, D.; Hall, E.; Ecclestone, K. Should we be using Learning Styles? *Learn. Ski. Res. Cent.* **2004**, 84.
21. Felder, R.M.; Spurlin, J. Applications, reliability and validity of the Index of Learning Styles. *Int. J. Eng. Educ.* **2005**, *21*, 103–112.
22. Felder, R.M.; Brent, R. Understanding student differences. *J. Eng. Educ.* **2005**, *94*, 57–72. [CrossRef]
23. Ramírez-Correa, P.E.; Rondan-Cataluña, F.J.; Arenas-Gaitán, J.; Alfaro-Perez, J.L. Moderating effect of learning styles on a learning management system's success. *Telemat. Inform.* **2017**, *34*, 272–286. [CrossRef]

24. Quinn, M.M.; Smith, T.; Kalmar, E.L.; Burgoon, J.M. What type of learner are your students? Preferred learning styles of undergraduate gross anatomy students according to the index of learning styles questionnaire. *Anat. Sci. Educ.* **2018**, *11*, 358–365. [CrossRef] [PubMed]
25. Quinlan, J.R. Induction of Decision Trees. *Mach. Learn.* **1986**, *1*, 81–106. [CrossRef]
26. Salzberg, S.L. C4.5: Programs for Machine Learning by J. Ross Quinlan. Morgan Kaufmann Publishers, Inc., 1993. *Mach. Learn.* **1994**, *16*, 235–240. [CrossRef]
27. Wu, X.; Kumar, V.; Ross, Q.J.; Ghosh, J.; Yang, Q.; Motoda, H.; McLachlan, G.J.; Ng, A.; Liu, B.; Yu, P.S.; et al. Top 10 algorithms in data mining. *Knowl. Inf. Syst.* **2008**, *14*, 1–37. [CrossRef]
28. Romero, E.; Villar, P.; Gómez-Fraguela, J.A.; López-Romero, L. Measuring personality traits with ultra-short scales: A study of the Ten Item Personality Inventory (TIPI) in a Spanish sample. *Pers. Individ. Dif.* **2012**, *53*, 289–293. [CrossRef]
29. Jena, R.K. Predicting students' learning style using learning analytics: A case study of business management students from India. *Behav. Inf. Technol.* **2018**, *37*, 978–992. [CrossRef]
30. Troussas, C.; Krouska, A.; Sgouropoulou, C.; Voyiatzis, I. Ensemble learning using fuzzy weights to improve learning style identification for adapted instructional routines. *Entropy* **2020**, *22*, 735. [CrossRef]
31. Carvalho, L.M.C.; do Freixo Pereira, J.M.; Dias, R.M.T.S.; Noronha, A.B. Learning Styles in Business Management students: An empirical study in Portuguese Higher Education. *Adm. Ensino Pesqui.* **2020**, *21*, 348–384. [CrossRef]
32. Pasina, I.; Bayram, G.; Labib, W.; Abdelhadi, A.; Nurunnabi, M. Clustering students into groups according to their learning style. *MethodsX* **2019**, *6*, 2189–2197. [CrossRef]
33. Kaufman, L.; Rousseeuw, P.J. *Finding Groups in Data: An Introduction to Cluster Analysis*; John Wiley & Sons: Hobeken, NJ, USA, 2009; Volume 344, ISBN 0470317485.
34. Tulsi, P.K.; Poonia, M.P.; Priya, A. Learning Styles of Engineering Students. *J. Eng. Educ. Transform. Citeseer* **2016**, *30*, 44. [CrossRef]
35. Vita, G. De Learning styles, culture and inclusive instruction in the multicultural classroom: A business and management perspective. *Innov. Educ. Teach. Int.* **2001**, *38*, 165–174. [CrossRef]
36. Buxeda, R.J.; Moore, D.A. Using learning-styles data to design a microbiology course. *J. Coll. Sci. Teach.* **1999**, *29*, 159.
37. Palou, E. Learning Styles of Mexican Food Science and Engineering Students. *J. Food Sci. Educ.* **2006**, *5*, 51–57. [CrossRef]
38. Seery, N.; Gaughran, W.F.; Waldmann, T. Multi-Modal Learning in Engineering Education. In Proceedings of the 2003 ASEE Conference and Exposition, Nashville, TN, USA, 22–25 June 2003; American Society for Engineering Education: Washington, DC, USA, 2003; p. 17.
39. Kuri, N.P.; Truzzi, O.M.S. Learning styles of freshmen engineering students. In Proceedings of the 2002 International Conference on Engineering Education, Manchester, UK, 18–21 August 2002; pp. 1–5.

applied
sciences

Article

Modeling E-Behaviour, Personality and Academic Performance with Machine Learning

Serepu Bill-William Seota [1,*], Richard Klein [1] and Terence van Zyl [2]

[1] School of Computer Science and Applied Mathematics, University of the Witwatersrand, Johannesburg 2001, South Africa; richard.klein@wits.ac.za
[2] Institute for Intelligent Systems, University of Johannesburg, Johannesburg 2092, South Africa; tvanzyl@uj.ac.za
* Correspondence: at830828@student.reading.ac.uk

Featured Application: The e-behaviour, personality and performance evaluation frameworks described in this article can be used by students and academic staff alike to monitor performance and online behaviour as it relates to performance. Being aware of e-behavioural patterns is a starting point to improving the academic performance of individual students and groups of students. The methodology can be used to inform the extent to which a course is to be adapted such that it encourages students to engage in behaviour that promotes better academic performance.

Abstract: The analysis of student performance involves data modelling that enables the formulation of hypotheses and insights about student behaviour and personality. We extract online behaviours as proxies to Extraversion and Conscientiousness, which have been proven to correlate with academic performance. The proxies of personalities we obtain yield significant ($p < 0.05$) population correlation coefficients for traits against grade—0.846 for Extraversion and 0.319 for Conscientiousness. Furthermore, we demonstrate that a student's e-behaviour and personality can be used with deep learning (LSTM) to predict and forecast whether a student is at risk of failing the year. Machine learning procedures followed in this report provide a methodology to timeously identify students who are likely to become at risk of poor academic performance. Using engineered online behaviour and personality features, we obtain a classification accuracy (κ) of students at risk of 0.51. Lastly, we show that we can design an intervention process using machine learning that supplements the existing performance analysis and intervention methods. The methodology presented in this article provides metrics that measure the factors that affect student performance and complement the existing performance evaluation and intervention systems in education.

Keywords: e-behaviour; big five personality; student performance

Citation: Seota, S.B.-W.; Klein, R.; van Zyl, T. Modeling E-Behaviour, Personality and Academic Performance with Machine Learning. *Appl. Sci.* **2021**, *11*, 10546. https://doi.org/10.3390/app112210546

Academic Editor: Juan Cruz-Benito

Received: 20 August 2021
Accepted: 3 November 2021
Published: 9 November 2021

Publisher's Note: MDPI stays neutral with regard to jurisdictional claims in published maps and institutional affiliations.

1. Introduction

The evaluation and analysis of the factors that affect the academic performance of tertiary students stem from a need to improve student throughputs. Richiţeanu-Năstase and Stăiculescu [1] identify several reasons why post-secondary educational institutions have a low rate of completion. They name three main reasons: first, a lack of support (such as academic counselling services), second, the student's background, and third, an inability to adapt to the curriculum.

In addressing student performance, we consider their grades at the end of a study programme as a measure of their performance. We also refer to performance as *risk* or *risk of failure* since an increase in performance results in a lowered risk of failure. The *e-behaviour* of a student is "a pattern of engagement with a Learning Management System (LMS)", and *personality* according to Wright and Taylor [2] refers to "[...] the relatively stable and enduring aspects of individuals which distinguish them from other people and form the basis of our predictions concerning their future behaviour".

Traditional approaches to revealing relationships between student behaviour, personality and performance include questionnaires, surveys and interviews. The reliability of a questionnaire would depend on the contextual framework for the research and the metric construct used for outcomes. However, respondents' biases from the above qualitative methods of data collection can compromise the accuracy of their responses [3–5]. Furthermore, it has not proven easy to measure the reliability of an opinion [6], especially for each individual in a population. We address the self-reporting problems using unobtrusive and automated approaches that measure how students behave rather than how they think they behave. For instance, instead of asking, 'In how many weekly online discussions do you participate?', we instead obtain the exact number of discussions from an LMS register. The models developed in this research use quantitative metrics to proxy behaviour and personality traits traditionally obtainable from surveys. These metrics are used to draw correlations with features later used to predict student performance. From e-behaviour and personality, we modelled an intervention framework that supplements current student intervention systems.

We extract behavioural insights linked to two of the five personality traits in the Big Five personality model through a quantitative analysis—Conscientiousness and Extraversion. We use statistical metrics to extract forum and login behaviours, respectively. We define the relationships between these metrics and online behaviours, detailing the relationships between a student's expressions of personality traits through their behaviours. Through this research, we:

1. Define a framework for personality traits and behaviour in the context of student online engagement;
2. Show the relationship between Bourdieu's Three Forms of Capital and academic performance;
3. Show the relationship between personalities and academic performance through e-behaviours;
4. Show that we can use e-behaviour and machine learning to predict student performance;
5. Highlight the importance of the explainability of modelled personality traits and e-behaviours.

This work contributes to the prediction of student outcomes using online behaviours in the following ways:

- We present a framework and methodology for arriving at predictive models for student performance starting with personality traits. These traits are the drivers of online behaviours that generate features that are predictive of performance.
- We argue for the use of online behaviours and proxies for the personality traits Conscientiousness and Extraversion.
- We demonstrate that online behaviours that are strongly associated with the identified personality traits correlate with student performance in a statistically significant way.

1.1. Literature Review

A standard psychological framework for measuring personality is the Five-Factor or OCEAN (Openness, Conscientiousness, Extraversion, Agreeableness and Neuroticism) model (Costa and McCrae [7], Poropat [8], Furnham et al. [9], Ciorbea and Pasarica [10], Kumari [11], Morris and Fritz [12]). Recent research by Morris and Fritz [12] has shown that Conscientiousness and Extraversion are significantly correlated with student educational outcomes. In this research, we build upon the vast body of literature that supports these two personality traits as being correlated with student performance [8–14].

A revised Neuroticism–Extraversion–Openness Personality Inventory (NEO PI-R) [15] expands each of the OCEAN personality trait's six facets. For Conscientiousness, these facets are competence, orderliness, dutifulness, achievement striving, self-discipline and deliberation. Activity, assertiveness, excitement seeking, gregariousness, positive emotion and warmth are the facets of Extraversion. Wilt and Revelle [16] define Extraversion as the 'disposition to engage in social behaviour', which links significantly to the gregariousness facet. *Gregariousness* is defined as the 'tendency for human beings to enjoy the company of others and to want to associate with them in social activities' [17]. *Dutifulness* is defined as the characteristic of being motivated by a sense of duty [18]. In this article, we model for the orderliness facet of Conscientiousness and the gregariousness facet of Extraversion.

Recent work by Akçapınar [19] and Huang et al. [20] has shown that the usage of online behaviours alone does not necessarily lead to features that are predictive of student performance. We argue that this may be due to the many features that can be engineered from the time series of log data representing online student behaviour. For instance, the Tsfresh library can extract over 40 time-series features. The link between online behaviours and the personalities that underlie them has not been extensively explored. We follow the argument of Khan et al. [21] and postulate that starting from a principled approach that is grounded in personality traits will lead to a more viable set of features and metrics.

Contribution to Existing Evaluation Systems

The University of Witwatersrand's (the University's) academic and student support staff members have access to three standard systems of identifying the likelihood of students completing a programme. These systems can be broadly grouped into grades, questionnaires, observing a student's grades for that programme over time, and one-on-one consultations by a counsellor or lecturer with the student.

Questionnaires have two significant limitations. Firstly, they are not offered throughout the teaching period and, secondly, they are anonymous, meaning there is no easy way of linking students at risk to their programmes. Fowler and Glorfeld [22], Poh and Smythe [23], Evans and Simkin [24] show that prior performance or grades are a reliable measure for future performance. By their high-touch nature, observing grades and consultations are usually not anonymous. The advantage of these two systems is that they give a detailed response to students' feelings towards their programmes and are, thus, potentially corrective (can help resolve poor performance). The disadvantages are that one-to-one consultations and grades are often retroactive rather than proactive and not conducted at scale or sufficiently continuously.

The limitations in gauging student performance through the above mechanisms give rise to a proposal for using e-behaviour machine learning models, while models that fit students' e-behaviour do not guarantee similar reliability, e-behaviour machine learning models have some advantages over grades, questionnaires and consultations. Table 1 shows a comparison of evaluation systems. Whether an evaluation system is continuously proactive, corrective, feasible at scale and reliable depends on:

- The contextual framework for the research;
- The metric construct used for measuring each of these variables and outcomes;
- How the variables are used in an academic setting.

In Table 1, note that e-behaviour models are the only system of evaluation that is continuously proactive—can be monitored at any point in time to take corrective action.

Table 1. Comparison of evaluation systems.

System	Continuously Proactive	Corrective	Easily Feasible at Scale	Reliable
Questionnaires	✗	✗	✓	✗
Previous Grades	✗	✗	✓	✓
Consultation	✗	✓	✗	✓
e-behaviour Models	✓	✗	✓	To be shown

1.2. Bourdieu's Three Forms of Capital and Student Success

Bourdieu's Three Forms of Capital is a framework that suggests that economic, cultural and social capital that an individual can leverage regulates their level of success. We use this framework to support our investigation of the economic, cultural and social capital that a student has available to them as each form of capital relates to their academic performance. Dauter [25] defines economic sociology as

> "[...] the application of sociological concepts and methods to the analysis of the production, distribution, exchange, and consumption of goods and services."

Economic sociology has been used extensively by Bourdieu and Richardson [26] who argue that an individual's possession of three forms of capital regulates their social positions and ability to access goods and services. These three forms of capital are economic capital, cultural capital and social capital. The Three Forms can be considered essential to a student obtaining good grades and acquiring the services they need to improve their grades. We refer to our proxies for economic and cultural capital as the *background* of a student.

1.2.1. Social Capital

Bourdieu and Richardson [26] define social capital as:

> "The aggregate of the actual or potential resources which are linked to the possession of a durable network of more or less institutionalised relationships of mutual acquaintance and recognition."

The above definition describes social capital as a resource that is available between people due to their relationships. An individual may accrue social capital by being part of relationships. Carpiano [27] uses the framework by Bourdieu and Richardson [26] to build onto the theory of social capital. Carpiano [27] categorises the social capital available to individuals into four types, namely: social support, social leverage, informal social control and community organisation participation.

The above four types of social capital are available to students, forming relationships for social or academic purposes. Hallinan and Smith [28] refer to these intra-cohort groups as *social networks* or *cliques*. The common saying, *show me your friends and I will show you your future*, is commonly used to describe the relationship between an individual's affiliates and their results. In this research, these results are referred to as their *Grade* or their *Outcome*. The hypothesis that a student has access to some *social capital* has been validated to various extents by Hallinan and Smith [28] and is also adopted in this research.

A limitation with the social capital frameworks by Bourdieu and Richardson [26], Carpiano [27] and Song [29] is that they provide no standard measure of social capital. The definition of social capital leaves no room for a well-defined metric. In our research, a student's social network is evidence of their social capital and is called their *Academic group*. In an academic setting, a student's *quality* of resources social capital can be defined in terms of the aggregate grades of their Academic group. The relationships between Academic groups and Grades is modelled in Sections 3.4 and 3.5.

1.2.2. Cultural Capital

According to Hayes [30], cultural capital is a set of non-economic factors that influence academic success, such as family background, social class and commitments to education, and do not include social capital. Bourdieu and Richardson [26] categorise cultural capital into three forms, namely:

1. Institutionalised cultural capital (highest degree of education);
2. Embodied cultural capital (values, skills, knowledge and tastes);
3. Objectified cultural capital (possession of cultural goods).

In this research, the features we selected in Section 2.5 are proxies of 1 and 2. Smith and White [31] found that success in obtaining a degree relates strongly to gender and ethnicity. Caldas and Bankston [32] found that students' cultural capital affects their performance.

1.2.3. Economic Capital

Bourdieu and Richardson [26] define *economic capital* as material assets that are 'immediately and directly convertible into money'. In turn, an individual's monetary leverage can be converted into cultural and social capital [26].

Bourdieu and Richardson [26] recognise that an individual can increase their social and cultural capital by making use of their economic capital. An individual who leverages their economic capital can obtain more resources to improve their cultural capital. For instance, an individual can improve their cultural capital through improvement in their position in society. By investing in formal or informal education beyond the classroom, a student may increase their knowledge and the amount of cultural capital available to them. Fan [33] observed that a student's quality and level of education was affected by their cultural and economic capital.

Section 3.1 reveals the relationships between student background (*background* refers to cultural and economic capital) and academic performance.

2. Methodology

2.1. Data Preprocessing

The data were composed of files with logs on the *Moodle* LMS database at our university for first-, second-and third-year students who were enrolled in Applied Mathematics and Computer Science modules in the 2018 academic year. After examining distributions and removing students who had no grade records, the data were reduced to time series patterns, aggregated by each day of the semester. The models were fitted on the open-semester logs recorded (from the beginning of the semester till two weeks before exams). The advantage of using only open-semester data was not only that it helped us understand the predictive power of the behaviour, but it also eliminated effects of sudden changes in behaviours that were forced upon students as examinations approached [12]. The target variable for all experiments was the aggregate Grade (out of 100 points) of online assessments, including examinations, that the student obtained over the year.

2.2. Importance and Choice of Personality Traits

The university LMS contained several tables that each provided different information. We checked each table's appropriateness in modelling any of the OCEAN traits. The Forums and Logins Tables contained logs with details about student interaction, and were, thus, chosen as primary tables from which to source behavioural information for our proxies for Conscientiousness and Extraversion. By comparison, Openness, Agreeableness and Neuroticism were more complex to model, given the available data and the lack of validation of a link to academic performance within the literature.

Our data linked closest to the dutifulness facet of Conscientiousness and the gregariousness facet of Extraversion. Alternative formulations of each trait were considered and are described in Section 5.1. We used quantitative proxies to model Conscientiousness (Dutifulness) and Extraversion (Gregariousness).

2.3. Encoding Personality Traits

According to Ajzen [34], Campbell [35], human behaviour can be explained by reference to stable underlying dispositions or personality. Wright and Taylor [2] define personality as:

> 'the relatively stable and enduring aspects of individuals which distinguish them from other people and form the basis of our predictions concerning their future behaviour'.

Therefore, Extraversion and Conscientiousness were modelled as single-valued averages that did not vary through time. Our choice to encode personality traits as unvarying values was based on the theory by Wright and Taylor [2], Ajzen [34], Campbell [35], Hemakumara and Ruslan [36].

We acknowledge that 'there is a complex relationship between personality and academic performance' [8]. As a result, the same complexities could be expected between our proxies of Conscientiousness, Extraversion and Performance. These relationships were not controlled for, since they were not present in our data. However, given the data and interaction between them, it was important to control for the variables, in light of research by Poropat [8].

Challenges against Encoding Personality Traits

The above definitions of personality and their link to behaviour may cause a belief that personality and behaviour should be measured identically. However, to understand the separate correlations between either personality and performance or e-behaviour and performance, it was essential to encode an individual's *stable aspects* (personality traits that are less likely to change) differently from their *changing* e-behaviour. As a result, personality metrics were aggregated while e-behaviour was modelled to vary over time.

2.4. Encoding Performance

Three measures of performance were constructed, namely, *Grade*, *Outcome* and derived *Safety Score*. Grade is a continuous label that indicates the mean of a student's performance across all modules taken. This label was continuous and ranged between 0.00 and 100.00. Outcome is a binary label that indicates whether a student obtained below 51 *Grade* points (*At-risk*) or at least 51 *Grade* points (*Safe*). That is, the Outcome was taken to measure the degree of *risk-of-failure*. Note that a *fail* was considered any grade below 50 Grade points. However, the boundary of 51 provided a buffer that allowed the models to reveal students who were close to failing (At-risk). Therefore, a student need not fail for them to be considered at risk. A student's Safety Score is a classification label used as a label of their predicted Outcome. A correct classification would assign a *Flagged* Safety Score for an At-risk student and an *Ignored* Safety Score for a student with a Safe Outcome.

Grade was used as a regressor against Extraversion level (Section 2.6) and Conscientiousness level (Section 2.9). *Outcome* was used as a label to the classification models in Sections 2.5 and 2.10.

2.5. Student Background

The raw Background Dataset consisted of 4748 students and 176 features on which experiments were conducted. These features captured answers by the student upon registration and data collected throughout their study—for instance, their high-school facilities, high-school subjects, age and city of residence. Table 2 shows a summary of the features after each phase of transformation.

The 169 categorical features were one-hot encoded, extending the number of features from 176 to 6623. Recursive Feature Elimination algorithm (RFE) with a Decision Tree was used to reduce the 6623 feature set's dimensionality.

Table 2. Background data feature count per phase of transformation.

Transformation Phase	Categorical Features	Total Features
Before One-hot	169	176
After One-hot	6616	6623
After RFE	5	5

Feature Selection Using RFE

RFE involved filtering through features with the lowest ranking of importance against Outcome, through the following procedure [37]:

1. Optimise the Decision Tree weights with respect to their objective function on a set of features, F;
2. Compute the ranking of importance for the features in F using the Decision Tree optimiser;

3. Prune the features with the lowest rankings from F;
4. Repeat 1–3 on the pruned set until the specified number of features is reached.

The RFE process produced five Background Features, explained with the Grade and Outcome variables in Table 3.

Table 3. Background data features and labels after RFE.

Feature	Description	Type	Values
Quintile	To which of the five categories a student's high-school belongs under the South African Government school standards; a 6 indicates private high-schools	Categorical	1–6
Gauteng Province	Whether a student completed their ultimate year of high-school at a school in *GP* (Gauteng Province)	Binary	No, Yes
Gender	Whether the student was female or male	Binary	Female, Male
Financial Assistance	Whether a student received financial aid from the National Student Financial Aid Scheme	Binary	No, Yes
Township School	Whether a student's high-school was situated in a township area	Binary	No, Yes
Grade (label)	Grade points out of 100 obtained, as defined in Section 2.4 on Encoding performance	Continuous	0.0–100.0
Outcome (label)	Risk of the student based on their Grade, as defined in Section 2.4 on Encoding performance	Binary	At-Risk, Safe

2.6. Extraversion and Academic Groups

Discussion, *Message* and *Time* independent variables, explained in Table 4, were used to engineer the Extraversion level (*Extraversion level* was the proxy for Extraversion) of a student, as well as formulate the Discussions and Collaboration groups.

Table 4. Forum table features.

Feature	Description	Type	Possible Values
Discussion	Discussion number. Messages that begin a topic and are posted as responses were assigned the same discussion number.	Categorical	0–337
Message	Contents of each forum post.	String	-
Time	Extracted from the *Created* variable. Indicates the time at which each *message* was posted.	yyyy-mm-dd	2018-01-05 to 2019-01-05
Grade (label)	Number of points out of 100 (Section 2.4).	Number	0.0–100.0

Forum Posts and Extraversion

The definition of social capital in Section 1.2 suggests that Extraversion or gregariousness can improve an individual's ability to accumulate social capital, which is correlated with academic performance. A way to model social interaction or gregariousness is by capturing the number of forum posts that an individual contributes to forum discussions. Hence, we chose the student's post count as a quantitative proxy for their *level* of Extraversion.

Each student was placed in an Extraversion-level group, E, representing the number of posts they contributed. Each level, E, was then assigned a mean Grade, G_E, computed by averaging the grades of all students in E. Table 5 shows the each Extraversion-level above its associated Grade.

Table 5. Input table–Extraversion level Grade against Extraversion level.

E	0	1	2	3	4	5	6	7	8	9	10	11	12	13	14
G_E	55.8	64.4	68.3	66.1	69.3	73.3	66.0	69.7	71.4	70.2	81.3	79.4	75.2	74.4	79.3

2.7. Student Discussions

A Discussion (group), d_i, was defined as any discussion that contains more than two students created on the Moodle LMS. A Discussion that had fewer than three students was not considered a Discussion by our definition. Linear OLS assumptions for Discussions containing only two or more students did not hold. Section 4.4 shows the reasons. Let $D = \{d_i\}_{i=1}^{k} = \{d_1, d_2, \ldots, d_k\}$ be the set of all Discussions, s_j represent any student who participated in discussion d_i, s_i represent a selected random student who participated in discussion d_i, $\mathbb{E}[Gd_i]$ represent the mean Grade of Discussion d_i and $G(s_j)$ is the grade of student s_j.

$$\mathbb{E}[Gd_i] = \frac{1}{n(d_i) - 1} \sum_{s_j \neq s_i} G(s_j), \tag{1}$$

where k represents the number of Discussions in D and $n(d_i)$ denotes the number of students in d_i. This section measured the correlation between $Gd_i(s_i)$ and $\mathbb{E}[Gd_i]$ by following Algorithm 1:

Algorithm 1: Correlation Between Mean Discussion Grade and Student Grade.

Result: $\hat{G}d_i(s_i) = \beta_0 \mathbb{E}[Gd_i] + \beta_1$

foreach $d_i \in D$ **do**

    Select s_i
    Obtain $Gd_i(s_i)$ from s_i
    Compute $\mathbb{E}[Gd_i]$
    Plot $(Gd_i(s_i), \mathbb{E}[Gd_i])$

end

Table 6 shows a sample of random student Grades against their Discussion's Grade Averages.

Table 6. Discussion table—random student's Grades against Discussion's Grade averages.

d_i	s_i	$\mathbb{E}[Gd_i]$	$Gd_i(s_i)$
0	23	83.08	92.75
1	728	56.81	59.25
2	833	49.75	48.50
⋮	⋮	⋮	⋮
336	79	75.35	90.75
337	15	74.87	70.25

2.8. Student Collaboration Groups

This section illustrates an alternative method to formulating an Academic Group, namely, the Collaboration group method. We correlated the Grades of students within each Collaboration group with the mean Grade of each Collaboration group.

The raw Forum Table was transformed into Table 7 below, which shows discussion participation per student. Each column, d_i, represents a discussion: 1 represents that the student participated in discussion d_i, while 0 shows that they did not participate in d_i.

Table 7. Sample table of Discussion participation.

s	d_0	d_1	d_2	d_3	d_4	...	d_{337}
1	0	0	0	0	0	...	1
2	1	1	0	0	0	...	0
3	0	0	1	1	0	...	0
...	...	...	...	...	...	...	...
1131	0	1	0	1	0	...	0
1132	1	0	0	0	0	...	0
1133	0	1	0	0	0	...	1

Let $C = \{c_i\}_{i=1}^{k} = \{c_1, c_2, \ldots, c_k\}$ be the set of all Collaboration groups. h_i is the *Host* of c_i with $H = \{h_i\}_{i=1}^{k} = \{h_1, h_2, \ldots, h_k\}$ being the set of all Hosts, one for each Collaboration group.

A Collaboration group, c_i, that was *hosted by* student, h_i, was defined as the group of more than two students with whom h_i shared at least one discussion. Any group with two or fewer students was not considered a Collaboration group by our definition; OLS relationships analogous to those in this section did not hold for groups containing only two or more students. The reasons were presented under Section 4.4. h_i may host a maximum of one Collaboration group. Let $\mathbb{E}[Gc_i]$ represent a the mean Grade of c_i, and $\mathbb{E}[Gc_i]_i(h_i)$ represent the Grade of h_i where $\mathbb{E}[Gc_i]$ represents the mean Grade of c_i which excludes $Gc_i(h_i)$, as in the case with $\mathbb{E}[Gd_i]$ and $Gd_i(s_i)$ in Equation (1).

The *k*NN algorithm was used to compute the Collaboration group for each student, using the **Collaboration group policy** specified in the below paragraph. By this policy, not all students fit the qualify to host a Collaboration group.

We designed the conditions necessary to define the Collaboration group policy; let h_* be a *candidate* Host of a Collaboration group, with c_* representing the Collaboration group to be hosted by student, h_*. $n(c_*)$ represents the number of students in c_*, s_1, s_2 and s_3 are any three students in the cohort, and h_i represents a (qualified) Host to their (unique) Collaboration group, c_i.

Collaboration group Policy: c_* becomes a Collaboration group, c_i, if and only if $n(c_*) > 2$ students. Equivalently, if h_* shares a discussion with s_1, s_2 and s_3, then h_* *qualifies* as a Host, h_i, and $c_i = \{s_1, s_2, s_3\}$. If $n(c_*) \leq 2$ students, then h_* remains a candidate until they share a discussion with at least one more member.

A sample set of the Hosts, h_i, and their Collaboration groups, c_i is shown in Table 8. Each entry in column c_i was a set of indices that represented students in c_i, while column $Gc_i(h_i)$ showed the Grades of the Hosts. $\mathbb{E}[Gc_i]$ represented the mean Grades of each c_i.

Table 8. Collaboration—groups and grades.

h_i	c_i	$Gc_i(h_i)$	$\mathbb{E}[Gc_i]$
1	{5, 48, 3, 138}	73.00	47.68
2	{119, 172, 199}	81.80	67.62
3	{40, 35, 20, 16, 51}	90.75	69.80
4	{90, 200, 28, 33, 94, 142, 42, 101, 84}	49.00	62.08
5	{81, 209, 143, 206, 12, 150}	98.25	63.04
6	{142, 33, 28, 42}	59.25	58.25
7	{65, 190, 107, 8, 173}	46.60	74.51

2.9. Logins and Conscientiousness

Section 2.2 explained the facets that describe each personality trait. Our model of Conscientiousness related closely with dutifulness. Barrick et al. [38] and Campbell [39] theorise that Conscientiousness is linked to an individual's choice to expend a level of effort. Therefore, modelling dutifulness required a formulation that captured the average logins per week that the student performed throughout the programme. This model of the Conscientiousness level captured the facet of dutifulness and the choice to expend effort (*Conscientiousness level* is the proxy for Conscientiousness). Let $C(s)$ be a variable that represents the Conscientiousness level of student s. $C(s)$ was modelled as the average number of logins over the period spanning a student's active weeks. For each student, $C(s)$ was formulated as:

$$C(s) = \frac{\sum_t^{17} L(s)_t}{W(s)}, \tag{2}$$

where $W(s)$ is the number of weeks spanned between the student's first and last login.

The reason for modelling $C(s)$ as the *average* number of active days per week instead of the *total* number of logins over the period was that the *average* normalised the data. Averaging reduced biases caused by differences in the number of days per cohort, per subject and programme, that students were expected to log in.

Each personality trait proxy was regressed against the students' grades for the semester using the Ordinary Least Squares (OLS) regression method [40]. The associated slope and correlation coefficients, p-values and Slope coefficient with 95% confidence intervals were reported.

To date, several longitudinal studies investigating academic performance and personality have used *effects*, *determines* or *predicts* to mean the *relationship* or *correlation* between personality traits and performance (for example, works by Poropat [8], Chamorro-Premuzic and Furnham [13], Blumberg and Pringle [41] and Morris and Fritz [12]. Without knowing the causality of personality traits on performance in our study, we adopted the same terminology for ease of reference and comparison.

2.10. Behaviour–Personality and Behaviour Model

The Behaviour–Personality model (B-PM) consisted of two components: the behavioural component was the Login Sequences of students, while the personality component augmented the Login Sequences. The traits that composed the personality component were the Extraversion and Conscientiousness levels. The Behaviour Model (BM) consisted of only the Login Sequences of students as input.

For each student s, we engineered the Login Sequence ($\{L(s)_t\}$), Extraversion level ($E(s)$) and Conscientiousness level ($C(s)$ by augmenting $\{E(s)_t\}_{t=1}^{17}$ and $\{C(s)\}_{t=1}^{17}$ as *sequences* of the *same* values that ran parallel to $\{L(s)_t\}_{t=1}^{17}$ through time t. This augmentation formed a 3×17 *input array of sequences*:

$$[\{L(s)_t\}, \{E(s)_t\}, \{C(s)_t\}],$$

where $\{L(s)_t\}_{t=1}^{17}$ is a sequence of values that *vary* through time t, $\{E(s)_t\}_{t=1}^{17}$ is a sequence of the *same* value through time t, so that $E(s)_t = E(s)_{t-1}$ for all Whole Numbers $t \in [2, 17]$, and $\{C(s)_t\}_{t=1}^{17}$ is a sequence of the *same* value through time t, so that $C(s)_t = C(s)_{t-1}$ for all Whole Numbers $t \in [2, 17]$. (See Table 9.)

This method of augmenting inputs in parallel was guided by its usage in Leontjeva and Kuzovkin [42]. As a result, The B-PM **input** for each student was the *array of sequences*:

$$[\{L(s)_t\}, \{E(s)_t\}, \{C(s)_t\}].$$

The B-PM **output** for each student was a Safety Score: *Flagged* for At-risk students, and *Ignored* for Safe students. These per student B-PM **input** and **output** structures are summarised in Table 9.

Table 9. B-PM training input and output summary.

Feature	Shape	Type	Example Value
$\{L(s)_t\}$ Login Sequence	(1×17)	A Sequence of Whole Numbers	$[3, 7, \ldots, 0]$
$\{E(s)_t\}$ Extraversion (Extraversion level Sequence) Sequence	(1×17)	A Sequence of Whole Numbers	$[8, 8, \ldots, 8]$
$\{C(s)_t\}$ Conscientiousness (Conscientiousness level Sequence) Sequence	(1×17)	A Sequence of Real Numbers	$[1.1, 1.1, \ldots, 1.1]$
Input: $[\{L(s)_t\}, \{E(s)_t\}, \{C(s)_t\}]$	(3×17)	An *Array of Sequences* of Real Numbers	$[[3, 7, \ldots, 0]$ $[8, 8, \ldots, 8]$ $[1.1, 1.1, \ldots, 1.1]]$
Output: Safety Score = $\{Flagged, Ignored\}$	(1×1)	Binary	Ignored

2.11. Algorithms for E-Behaviour, Personality and Performance Analysis

2.11.1. Decision Tree Classifier

The Decision Tree Classifier (DTC) is a supervised learning algorithm that iteratively assesses conditions on the values of features in a dataset to perform classification. DTC breaks down a decision-making process into a collection of simpler decisions, providing classifications that are easier to interpret than other statistical and machine learning models [43].

DTC Architecture

DTC was assembled from a root node, edges, internal nodes and leaf nodes. At the root node, DTC conducted a test on each observation's value. Based on its value, the root node assigned a resolution represented by an edge, which the observation then traversed. At the end of the traversed edge was an internal node. An example of a node's test was 'Gender?', and an example of an edge was 'Female'. This decision process continued through the rest of the internal nodes until the tree reached a leaf node, where a classification was determined. See Mitchell [44] for details on the DTC architecture.

Gini Impurity Index—Decision Factors

During prediction, an observation was predicted as part of a class after being checked through a series of conditions. An optimal decision tree resulted in an *optimal split*. An optimal split was achieved when each leaf node had the fewest possible train-set misclassifications (lowest impurity), and the tree had not been overfitted. Entropy and Gini Impurity Index are two commonly used metrics for impurity. The Gini Impurity Index (Gini) measures the relative frequency that a randomly chosen element from that set would be mislabelled. A Gini score greater than zero describes a node that contains samples belonging to different classes. Raileanu and Stoffel [45] suggest that the difference between Entropy and Gini is trivial. This research used Gini, which was interpretable.

Gini Calculation

The Gini value decreased as a traversal was determined down the tree towards its leaf nodes. The decrease happened as each internal node's condition aimed to separate the classes according to a criterion that resulted in more homogeneous separations and higher accuracy in the training data. However, as with other predictive models, a high training-set accuracy was generated at the risk of overfitting. A larger tree (with more edges and branches) was more likely to overfit than a smaller tree and could result in a Gini of 0 at the tree's leaf nodes. A Gini of 0 represented the minimum probability of misclassification over the training set, but could result in weak generalisability over the test set. Therefore, smaller trees were preferred to larger trees [44].

The Gini impurity index was calculated using the formula:

$$Gini = 1 - \sum_i p_i^2 \tag{3}$$

where p_i is the probability of class i.

Khalaf et al. [46] model DTCs on survey questions and answers that cover health, social activity and relationships of students to predict their academic performance. Topîrceanu and Grosseck [47] and Kolo et al. [48] provide literature in educational data mining and advocate for the use of the DTC due to its low complexity (with a run time of $O(m \times n \times log(n))$ and high interpretability. In Section 3.1, the DTC was used to select student economic and social capital features and predict the student Outcomes.

2.11.2. Ordinary Least Squares Linear Regression Analysis

Ordinary Least Squares (OLS) Linear Regression is a statistical model that estimates the linear relationship between one or more independent variables (regressors) and a dependent variable (regressand) [49]. Throughout this research, only one independent variable was used per regression model. Using one independent variable per model isolated the effect of each variable on Grade. A Regression model with one independent variable was called a Simple OLS Regression model. Each estimated or predicted value, $\hat{y}_i$, derived from the line of best-fit shown in Equation (5), could be determined by:

$$\hat{y}_i = \beta_0 x_i + \beta_1 + \epsilon_i, \tag{4}$$

where $\hat{y}_i$ is the predicted value of the ith independent variable, x_i. β_0 is the estimated slope coefficient of the model, representing the average marginal change in $\hat{y}_i$ for a unit increase in x_i. β_1, the fitted intercept of the model, represents the expected value of $\hat{y}_i$ when $x_i = 0$. $\epsilon_i \in \mathbb{R}$ is the residual term.

Every observed value, y_i, had an associated estimate or prediction value, $\hat{y}_i$. The line of best-fit,

$$\hat{y} = \beta_0 x + \beta_1, \tag{5}$$

was obtained by minimising the sum of the squares in the difference between the observed and predicted values of the dependent variable,

$$\sum(y_i - \hat{y}_i)^2 = (y_i - (\beta_0 x_i + \beta_1))^2. \tag{6}$$

The (linear) correlation coefficient between x and y was represented by r or $r(x,y)$. $r(x,y)$ measured the extent to which the independent variable, x, was correlated with the dependent variable, y. That is, $r(x,y)$ measured the degree of closeness of all points, (x,y), to the line of best-fit, $\hat{y} = \beta_0 x + \beta_1$. The correlation coefficient laid between -1 and 1, where a r of 1 or -1 meant that the change in y was directly proportional to the increase in x. In that case, x and y were stated to be perfectly correlated. That is,

$$r = 1 \implies \frac{y_i - y_{i+1}}{(x_i + 1) - x_i} = c, \tag{7}$$

for all values of i where x_i and y_i were defined, and where $c \in \mathbb{R}$. r was computed by:

$$r = \frac{\sum(x_i - \bar{x})(y_i - \bar{y})}{\sqrt{\sum(x_i - \bar{x})^2 \sum(y_i - \bar{y})^2}}, \tag{8}$$

where $\bar{x}$ represents the mean average of independent variable x and $\bar{y}$ represents the mean average of dependent variable, y.

The p associated with β_0 showed the probability of a hypothetical value, β_0^*, having an absolute value, β_0^*, that was at least as high as the observed β_0 by chance. The level of significance, α, was used as a threshold for a permissible p. In the domain relating to e-behaviour, personality and performance, the common α used was 0.05, or a 5% level of significance. In our regression models, β_0 was accompanied by a ($1 - \alpha = 95\%$) confidence interval, $\beta_0 \pm v$. Suppose $p < \alpha = 0.05$ for β_0. This meant that from 100 experiments on similar sample distributions, fewer than 5 experiments would produce a β_0^* value that laid outside of $\beta_0 \pm v$. Such a result meant that the regressor and regressand had a statistically significant correlation different from zero [49]. Statistical insignificance could indicate that x on its own did not yield reliable estimates, $\hat{y}$.

Statistical significance was important in analysing a student cohort's behaviour, since statistical significance confirmed the existence of a statistical relationship. Empirical significance refers to the magnitude of β_0 [49] and was also a measure of the model's practical value. One could be more confident in practical decisions if the relationship was not generated by chance (if the relationship was statistically significant). This *chance* was measured by the p.

2.11.3. Validity of OLS Regression Models

The data given in a model had to satisfy five OLS regression assumptions [49], namely:

1. Normality of model residuals. The residual for each point was given by $y_i - \hat{y}_i$. $s^2 + k^2$ was computed for the residuals, where s is the z-score returned by the test for skewness and k is the z-score returned by the test for kurtosis.
2. Residual Independence or lack of Autocorrelation in Residuals.
3. Linearity in Parameters.
4. Homoscedasticity of Residuals.
5. Zero Conditional Mean.
6. No Multicollinearity in Independent Variables.

A linear relationship that violated the OLS assumptions was not fit for an OLS model. Therefore, we constructed only OLS relationships that satisfied the assumptions. Mentioned were experiments where the OLS assumptions were violated. Linearity in Parameters, Homoscedasticity of Residuals and Zero Conditional Mean were verified for all Regression experiments whose results were analysed. The No Multicollinearity assumption was not verified since all OLS regression experiments were Simple. See Gujarati and Porter [49] for further details on the formulation of the OLS Regression model.

2.12. Long Short-Term Memory

The Long Short-Term Memory algorithm (LSTM) is a deep-learning architecture designed to model sequences for prediction [50]. The LSTM has been used in studies that range from predicting weather-induced background radiation fluctuation by Liu and Sullivan [50], to human motion classification and recognition by Wang et al. [51].

The backpropagation through time algorithm computes the error, $\mathbf{E}_t$, at every time step, t, and, then, computes the total error. The LSTM's parameters were updated to minimise the total error $\frac{\partial \mathbf{E}}{\partial \mathbf{W}}$ with respect to a weight parameter $\mathbf{W}$:

$$\frac{\partial \mathbf{E}}{\partial \mathbf{W}} = \sum_{t=1}^{T} \frac{\partial \mathbf{E}_t}{\partial \mathbf{W}}. \tag{9}$$

Letting $\mathbf{y_t}$ represent the output at time t, $\mathbf{h_t}$ represents the hidden state at time t and by applying the chain rule to the Recurrent Neural Network model, the total error in Equation (9) became:

$$\frac{\partial \mathbf{E}}{\partial \mathbf{W}} = \sum_{t=1}^{T} \frac{\partial \mathbf{E}}{\partial \mathbf{y}_t} \frac{\partial \mathbf{y}_t}{\partial \mathbf{h}_t} \frac{\partial \mathbf{h}_t}{\partial \mathbf{h}_k} \frac{\partial \mathbf{h}_k}{\partial \mathbf{W}}, \tag{10}$$

where $\frac{\partial \mathbf{h}_t}{\partial \mathbf{h}_k}$ involves a product of Jacobian matrices:

$$\frac{\partial \mathbf{h}_t}{\partial \mathbf{h}_k} = \frac{\partial \mathbf{h}_t}{\partial \mathbf{h}_{t-1}} \frac{\partial \mathbf{h}_{t-1}}{\partial \mathbf{h}_{t-2}} \cdots \frac{\partial \mathbf{h}_{k+1}}{\partial \mathbf{h}_k}. \tag{11}$$

Equation (11) illustrates the problem of vanishing gradients in Equation (9): when the gradient became progressively smaller as k increased, the parameter updates became insignificant

LSTMs are an architecture of Recurrent Neural Networks (RNNs). Bengio et al. [52] suggest that RNNs are challenging to train because of the vanishing error gradient problem. The following section stipulates how the LSTM's architecture mitigates the vanishing error gradient issue through LSTM cells that maintain a state $\mathbf{c}_t$ at every iteration t. The cell state $\mathbf{c}_t$ serves to *remember* and propagate cell outputs between time steps. Each cell state then allows for temporal information to become available in the next time step, adding greater context to the inputs $\mathbf{x}_t$ that follow.

The activation $\mathbf{h}_t$ of an LSTM unit was:

$$\mathbf{h}_t = \mathbf{o}_t \tanh(\mathbf{c}_t), \tag{12}$$

where

$$\mathbf{o}_t = \sigma\left(\mathbf{W}_{xo}\mathbf{X}_t + \mathbf{W}_{ho}\mathbf{h}_{t-1} + \mathbf{b}_o\right), \tag{13}$$

is an output gate that mitigates the amount of content in the memory to expose to the following time step and $\sigma : \mathbb{R} \to (0,1)$ is the logistic sigmoid function.

Given new memory content,

$$\mathbf{i}_t \tanh\left(\mathbf{W}_{xc}\mathbf{X}_t + \mathbf{W}_{hc}\mathbf{h}_{t-1} + \mathbf{b}_c\right), \tag{14}$$

where $\mathbf{i}_t$ represents the degree to which new memory is added to the memory cell, and is specified by an input gate

$$\mathbf{i}_t = \sigma\left(\mathbf{W}_{xi}\mathbf{X}_t + \mathbf{W}_{hi}\mathbf{h}_{t-1} + \mathbf{b}_i\right), \tag{15}$$

the cell state,

$$\mathbf{c}_t = \mathbf{f}_t\mathbf{c}_{t-1} + \mathbf{i}_t \tanh\left(\mathbf{W}_{xc}\mathbf{X}_t + \mathbf{W}_{hc}\mathbf{h}_{t-1} + \mathbf{b}_c\right), \tag{16}$$

could be updated by taking into account the previous cell state $\mathbf{c}_{t-1}$ and a term defined by the forget gate,

$$\mathbf{f}_t = \sigma\left(\mathbf{W}_{xf}\mathbf{X}_t + \mathbf{W}_{hf}\mathbf{h}_{t-1} + \mathbf{b}_f\right). \tag{17}$$

Consolidating Equations (12) to (17), the system of equations that describe each LSTM unit given by:

$$\mathbf{f}_t = \sigma\left(\mathbf{W}_{xf}\mathbf{X}_t + \mathbf{W}_{hf}\mathbf{h}_{t-1} + \mathbf{b}_f\right), \tag{18}$$

$$\mathbf{i}_t = \sigma\left(\mathbf{W}_{xi}\mathbf{X}_t + \mathbf{W}_{hi}\mathbf{h}_{t-1} + \mathbf{b}_i\right), \tag{19}$$

$$\mathbf{c}_t = \mathbf{f}_t\mathbf{c}_{t-1} + \mathbf{i}_t \tanh\left(\mathbf{W}_{xc}\mathbf{X}_t + \mathbf{W}_{hc}\mathbf{h}_{t-1} + \mathbf{b}_c\right), \tag{20}$$

$$\mathbf{o}_t = \sigma\left(\mathbf{W}_{xo}\mathbf{X}_t + \mathbf{W}_{ho}\mathbf{h}_{t-1} + \mathbf{b}_o\right), \tag{21}$$

$$\text{and} \tag{22}$$

$$\mathbf{h}_t = \mathbf{o}_t \tanh(\mathbf{c}_t). \tag{23}$$

Let B denote the input batch size (number of time stamps per input chunk), H denote the LSTM hidden state capacity, and D represent the dimensions of the inputs to the LSTM. Then, in Equations (18) through (23):

$$\mathbf{x}_t, \mathbf{h}_{t-1} \in \mathbb{R}^{BxD}, \tag{24}$$

$$\mathbf{f}_t, \mathbf{i}_t, \mathbf{c}_t, \mathbf{o}_t, \mathbf{h}_t \in \mathbb{R}^{BxH}, \tag{25}$$

$$\mathbf{W}_{xf}, \mathbf{W}_{xi}, \mathbf{W}_{xc}, \mathbf{W}_{xo} \in \mathbb{R}^{DxH}, \tag{26}$$

$$\mathbf{W}_{hf}, \mathbf{W}_{hi}, \mathbf{W}_{hc}, \mathbf{W}_{ho} \in \mathbb{R}^{H^2}, \text{ and} \tag{27}$$

$$\mathbf{b}_f, \mathbf{b}_i, \mathbf{b}_c, \mathbf{b}_o \in \mathbb{R}^{BxH}. \tag{28}$$

For an illustration, refer to Figure 1. The LSTM had four main gates that responded to the values of four functions determined by $\mathbf{f}, \mathbf{i}, \mathbf{c}$ and $\mathbf{o}$, represented in Equations (18) through (21). With the input data matrix $\mathbf{x}_t$ (data vector if $B = 1$) concatenated with previous output matrix $\mathbf{h}_{t-1}$ (vector if $B = 1$), the flow of inputs and outputs from the various gates described in the LSTM equations interacted as follows:

1. $\mathbf{h}_{t-1}$ and $\mathbf{X}_t$ were fed into the gate (or function) $\mathbf{f}$, where the output $\mathbf{f}_t$ laid in the open interval $(0, 1)$. $\mathbf{f}_t$ then interacted with previous cell state $\mathbf{c}_{t-1}$ through element-wise multiplication $\otimes$; thus, $\mathbf{c}_{t-1}$ held an interim cell state, $\mathbf{f}_t\mathbf{c}_{t-1}$. At this stage, $\mathbf{f}_t\mathbf{c}_{t-1}$ represented a state that had forgotten some previous cell state data in $\mathbf{c}_{t-1}$ that were captured as unimportant (note that importance was regulated by weight coefficients that were trained and stored in their respective weight matrices).

2. Whereas the forget gate $\mathbf{f}_t$ focused on regulating the extent to which previous data were forgotten, the input gate $\mathbf{i}_t$ focused on adding new data, scaled by their importance, or extent to which data should be added from the matrix comprised of $\mathbf{h}_{t-1}$ and $\mathbf{X}_t$.

3. The *tanh* gate obtained $\mathbf{h}_{t-1}$ and $\mathbf{X}_t$, but used the hyperbolic tangent *tanh* function to compute its outputs (between -1 and 1).

4. The result given by *tanh* and $\mathbf{i}_t$ was then multiplied element-wise and further added ($\oplus$) to $\mathbf{f}_t\mathbf{c}_{t-1}$, giving $\mathbf{c}_t$, shown in Equation (20).

5. The output gate $\mathbf{o}_t$ decided what values to output, given $\mathbf{h}_{t-1}$ and $\mathbf{X}_t$, and also computed its exposure to the following cell state based on trained importance.

6. Finally, the values of the cell state, $\mathbf{c}_t$, were passed through a *tanh* function and multiplied by the output gate result, $\mathbf{o}_t$, such that the LSTM unit kept only the output that it accounted for as important in $\mathbf{h}_t$, described by Equation (23).

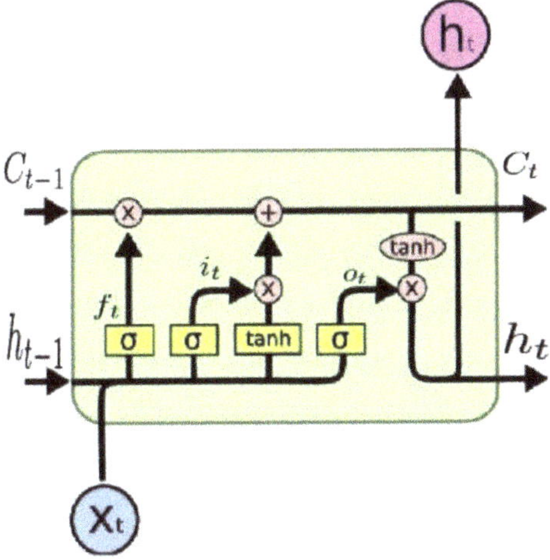

Figure 1. The LSTM unit kept a cell state throughout its operations, which served as input in the next time step. It also output $\mathbf{h}_t$, which supplemented the input $\mathbf{X}_t$ in the following time step. From Olah [53].

LSTM Problem Design

Let B (different from the B above) represent a $n \times T$ matrix containing the n number of e-behaviour sequences of all students. T is the length of each student's e-behaviour sequence. Let $B(s)$ represent a $1 \times T$ variable representing the e-behaviour sequence of student s, and let $B(s)_t$ be a scalar representing the value of $B(s)$ at time t. The LSTM learnt the interdependencies between variables B and $B(s)$ with the aim of classifying the risk (or determining the Safety Score) of s given the values of B and $B(s)$. That is, B and $B(s)$ were predictors of the Safety Score *(classification)* of s. Without loss of generality, this framework was used to predict the Safety Score of all students in Sections 3.1, 3.6 and 3.7.

2.13. Evaluation Metrics for Student Risk Classification

The Results Summary in Table 10 was used as an evaluation template for the classification problems in Sections 3.1, 3.6 and 3.7.

Results Summary

Table 10. Format of the Summary of Results.

		Safety Score (Prediction)		
		Flagged	**Ignored**	**Total**
Outcome	**At-risk**	a	b	$a+b$
(True Label)	Safe	c	d	$c+d$
	Total	$a+c$	$b+d$	$N = a+b+c+d$
	Outcome	**Precision**	**Recall**	
	At-risk	$\frac{a}{a+c}$	$\frac{a}{a+b}$	
	Safe	$\frac{d}{d+b}$	$\frac{d}{d+c}$	

The best-case scenario (where the e-behaviour model obtained a 100% accuracy) occurred where all students with an At-risk Outcome label were Flagged, and all students with a Safe Outcome label received an Ignored Safety Score.

Given an Outcome, At-risk, precision measured the proportion of students who were correctly Flagged as At-risk, a, against the total number of Flagged students. Precision was calculated as $\frac{a}{a+c}$, where c represented the number of students who should have been Ignored as Safe. Recall measured the proportion of students who were correctly Flagged as At-risk, a, against all At-risk observations, $\frac{a}{a+b}$. The same calculation generalised to the Safe Outcome.

It is essential to know the most important metrics to measure when evaluating a classifier's performance. Consider the Results Summary in Table 10. A perfectly accurate model resulted in a b and c equal to 0. The precision and recall scores would be 1 for both the At-risk and Safe Outcomes. None of our models achieved perfect accuracy—they conducted trade-offs regarding precision and recall. For a model whose objective was to classify all students who were at risk of failing, higher precision-recall scores for the At-risk Outcome werepreferred over higher precision-recall scores for the Safe Outcome. Furthermore, maximising the recall of the At-risk Outcome, $\frac{a}{a+b}$ (where the classifier recalled all students who were at risk) was preferred to maximising either the precision of the At-risk Outcome or the precision-recall scores of Safe students. Recall-maximisation would likely cause a low precision for the At-risk Outcome class (a high c). In such a case, however, no student who was At-risk would have been incorrectly Ignored.

The Overall Accuracy of a Model

While precision and recall are important metrics to measure a binary classifier's performance, they represent four different views of accuracy that must be analysed separately (precision and recall for At-risk and Safe Outcomes). When evaluating a model or accuracy, it is useful to obtain a single metric. A widely-used *accuracy* measure calculates the ratio between the correctly classified number of observations and the total number of observations. This *accuracy* is a good measure for balanced data, not for imbalanced data. For instance, if a test dataset contains 100 observations with an At-risk:Safe split of 10:90, a classifier can obtain an accuracy of 90% by classifying all students as Safe. An accuracy measure that combines the harmonic mean of precision and recall of either class is the *f-1 score*, whose effectiveness is surveyed by Hand and Christen [54]. The f-1 score produces two metrics (one for each Outcome) and does not concisely summarise the model's accuracy. By contrast, *Cohen's Kappa*, κ [55] is a metric that captures accuracy with a single value. The formula,

$$\kappa = (p_o - p_e)/(1 - p_e), \tag{29}$$

measures the *agreement* between the *predicted* Safety Score and the *true* Outcome. Landis and Koch [56] suggest using the scale in Table 11 to interpret the significance of κ values. In Identity 29, the observed accuracy (ratio between correctly classified number of students and total students), p_o, was adjusted for the expected *accuracy* when the classifier assigned a label randomly, p_e. In the example above, $p_e = 0.90$ and $p_o = 0.90$, giving a κ value of 0.00, or *no agreement* between the Outcomes and the Safety Scores assigned by the classifier. κ was, thus, more representative of a model's performance than the *accuracy* commonly used for data with balanced labels. *Chance* was an event that occurred when a classifier failed to fit an optimised objective function or had not learned anything from the data. In the above example, the κ of 0.00 signified that the classifier performed no better than chance.

Table 11. Cohen's Kappa interpretation.

κ	Level of Agreement
<0.00	Worse than chance
0.00–0.20	*Slight* agreement
0.21–0.40	*Fair* agreement
0.41–0.60	*Moderate* agreement
0.61–0.80	*Substantial* agreement
0.81–1.00	*Near-perfect* agreement

3. Results

3.1. Background Data and Grade

Figure 2 shows the linear correlation between the five Background predictor variables chosen with *Scikit-Learn*'s Recursive Feature Elimination algorithm and the Grade target variable, with a Decision Tree as its optimiser.

Quintile (r = 0.170) had the strongest linear correlation with Grade, followed by Township School ($r = -0.140$). The Background data's correlations indicate that higher Quintile high-schools generally performed better than students from lower Quintile schools. Students from Township schools performed worse than students from other schools.

Understanding that a relationship exists between the chosen features and Grade showed that these features caould inform the student's Grade and Outcome.

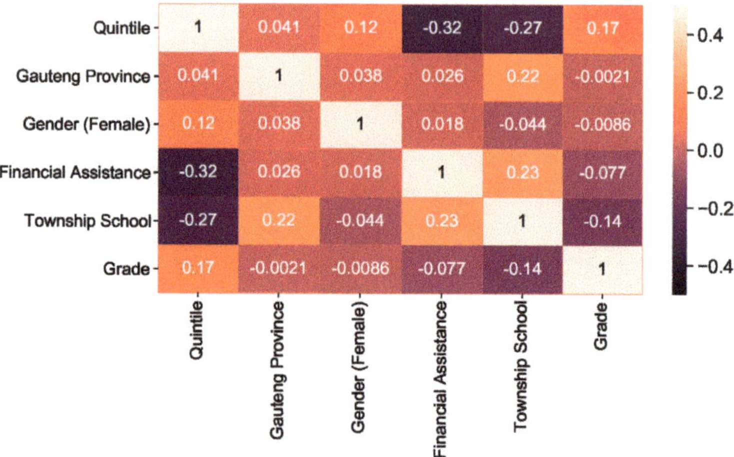

Figure 2. Pearson Correlation Coefficients of the Chosen Features. Quintile and Township School had the highest correlation with Grade.

Classifying a Student Based on Background Data

The classifier used was the Decision Tree Classifier. The train-set contained 3798, while the test-set contained 950 students. The train-test split was stratified by the Outcome of the students. A grid search on the train-set suggested a maximum tree depth of six and eight maximum leaves for the Decision Tree as presented in Table 12.

Table 12. Confusion matrix and summary of Background–Grade test set results.

		Safety Score (Prediction)		
		Flagged	**Ignored**	**Total**
Outcome (True Label)	At-risk	107	157	264
	Safe	153	533	686
	Total	260	690	950
	Outcome	**Precision**	**Recall**	
	At-risk	0.41	0.41	
	Safe	0.77	0.78	
	$\kappa = 0.18$			

If we refer to Table 12, we noted that 640 out of the 950 test observations were classified correctly, producing a κ of 0.18 (slight agreement) between the Safety Score and the Outcome. The precision score for the *Flag* students suggested that 107 of the 260 Flagged students were correctly Flagged; the remaining 153 were meant to be Ignored.

3.2. Extraversion-Level and Grade

Although the increase in the mean Grade with Extraversion level was apparent from the line of best-fit, this claim was confirmed by the OLS Regression model's output in Table 13. The corresponding plot is given in Figure 3. This result showed that students in higher Extraversion levels tended to achieve higher Grades, on average. The fit described in OLS Summary Table 13 showed a linear relationship, $\hat{G}_E = 1.269E + 62.422$. The *p*-values of 0.000 signified that $\hat{G}_E = 1.269f + 62.422$ was not a relationship by chance. Furthermore, the high r of 0.846 signified that G_E moved closely with E and could be inferred from E with a 95% confidence that $\beta_0 \in [0.771, 1.767]$.

Extraversion levels were ordinal, with each level indicating the number of posts in that level. Therefore, an E of one was a *lower* Extraversion level than an E of two.

Table 13. OLS Regression summary—Extraversion level Grade against Extraversion level.

Linear Equation: $\hat{G}_E = 1.269E + 62.422$				
Feature	**Coeff.**	**r**	**p-Value**	**Coeff. 95% CI**
E	1.269	0.846	0.000	[0.771, 1.767]
Intercept	62.422		0.000	[58.354, 66.491]

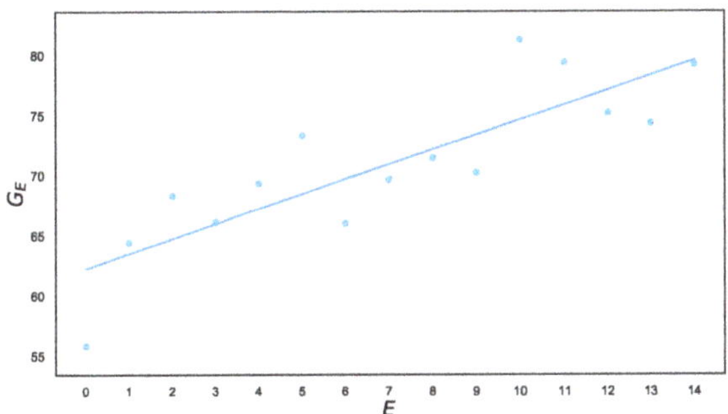

Figure 3. Extraversion-level Grade against Extraversion-level. $r = 0.846$, $p = 0.000$.

3.3. Conscientiousness-Level and Grade

There was a positive relationship between $C(s)$ and $G(s)$, with a β_0 coefficient p-value of 0.000 as supported by Table 14 and its corresponding plot in Figure 4. An increase of 1 in a student's Conscientiousness level corresponded to an average increase of 5.988 Grade points out of 100.

Table 14. OLS regression summary—average number of weekly active days against Grade.

Linear Equation: $\hat{G}(s) = 5.988C(s) + 39.829$				
Feature	**Coeff.**	**r**	**p-Value**	**Coeff. 95% CI**
$C(s)$	5.988	0.319	0.000	[4.129, 7.847]
Intercept	39.829		0.000	[34.426, 45.232]

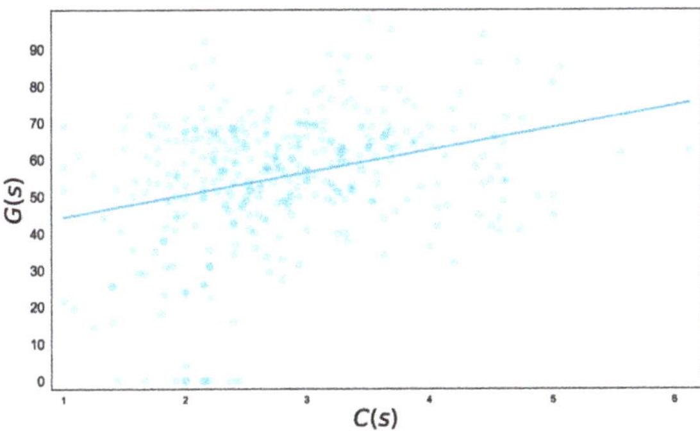

Figure 4. Average Number of Weekly Active Days against Grade. $r = 0.319$, $p = 0.000$.

3.4. Student Discussions and Grade

The OLS Regression Summary for the linear relationship between $\mathbb{E}[Gd_i]$ and $Gd_i(s_i)$ in Table 15 showed that the linear equation of the line of best-fit was given by $\hat{G}d_i(s_i) = 0.528\mathbb{E}[Gd_i] + 35.607$. The corresponding plot is given in Figure 5. The β_0 coefficient of $\mathbb{E}[Gd_i]$ and its statistical significance ($p = 0.011$) indicated that the a marginal increase in $\mathbb{E}[Gd_i]$ of one Grade point corresponded to an average increase of 0.528 in $Gd_i(s_i)$. The $r(\mathbb{E}[Gd_i], Gd_i(s_i))$ of 0.421 indicated a strong correlation between the mean Grade of a Discussion ($\mathbb{E}[Gd_i]$) and the Grade of a student, ($Gd_i(s_i)$), chosen at random who participated in that Discussion. This correlation also held for any other set of randomly selected students.

Table 15. OLS regression—random student's Grades against Discussion's Grade averages.

Linear Equation: $\hat{G}d_i(s_i) = 0.528\mathbb{E}[Gd_i] + 35.607$				
Feature	**Coeff.**	**r**	**p-Value**	**Coeff. 95% CI**
$\mathbb{E}[Gd_i]$	0.528	0.421	0.011	$(0.131, 0.925)$
Intercept	35.607		0.014	$(7.789, 63.425)$

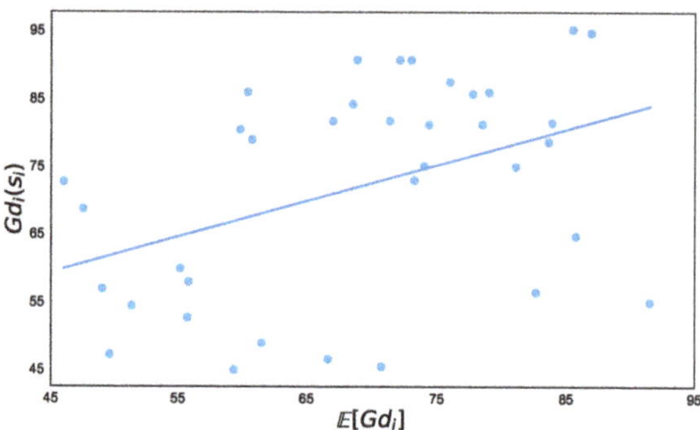

Figure 5. Random Student's Grades against Discussion's Grade Averages. $r = 0.421$, $p = 0.011$.

3.5. Student Collaboration—Groups and Grade

Table 16 shows the OLS regression results of the fit between $Gc_i(h_i)$ and $\mathbb{E}[Gc_i]$. The corresponding plot is given in Figure 6.

The coefficient of $\mathbb{E}[Gc_i]$ in Table 16 showed a marginal increase in $\mathbb{E}[Gc_i]$ of one Grade point corresponded to an estimated increase of 0.984 in $Gc_i(h_i)$. The $r(\mathbb{E}[Gc_i], Gc_i(h_i))$ of 0.479 indicates a strong correlation between the average Grade of a Collaboration group—$\mathbb{E}[Gc_i]$—and the Grade of its Host—$Gc_i(h_i)$.

Table 16. OLS regression summary—random student's Grades against Collaboration group's Grade averages.

Linear Equation: $\hat{G}c_i(h_i) = 0.984\mathbb{E}[Gc_i] + 5.175$				
Feature	**Coeff.**	**r**	**p-Value**	**Coeff. 95% CI**
$\mathbb{E}[Gc_i]$	0.984	0.479	0.004	$(0.334, 1.663)$
Intercept	5.175		0.797	$(-35.501, 0.975)$

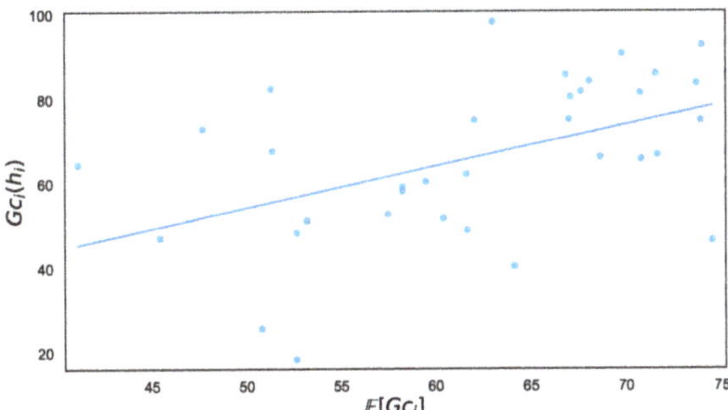

Figure 6. Random-Student's Grades against Collaboration-group's Grade Averages. $r = 0.479$, $p = 0.004$.

3.6. B-PM and Outcome

Table 17 shows B-PM's results. The κ of 0.51 showed a moderate agreement between B-PM's predicted Safety Scores and actual student Outcomes. B-PM's precision for the At-risk Outcome group showed that out of the 39 Flagged students, 27 were Flagged correctly (since they ended up at risk of failing). A total of 12 out of the 39 Flagged students were not meant to be Flagged. The At-risk recall indicated that out of 46 At-risk students, 27 were correctly Flagged, and the remaining 19 were incorrectly Ignored.

B-PM performed better at classifying Safe students than at classifying At-risk students: only 12 out of 124 Safe students were incorrectly Flagged, and 19 out of 131 Ignored students were wrongly Ignored.

Table 17. Confusion matrix and summary of B-PM test set results.

		Safety Score (Prediction)		
		Flagged	Ignored	Total
Outcome	At-risk	27	19	46
(True Label)	Safe	12	112	124
	Total	39	131	**170**
	Outcome	**Precision**	**Recall**	
	At-risk	0.69	0.59	
	Safe	0.85	0.90	
		$\kappa = 0.51$		

3.7. BM and Outcome

This section reports on the results of a modified model of B-PM without the $\{E(s)_t\}$ and $\{C(s)_t\}$ input Sequences. The comparison helped determine the change in the accuracy of B-PM after removing its personality components. This resulting model was called the Behaviour Model (BM); the only difference between BM and B-PM is that BM only has one input Sequence, $\{L(s)_t\}$.

Table 18 shows BM's results. For reference, the comparable B-PM results were shown in brackets. The At-risk recall of BM equalled the At-risk recall of B-PM, meaning that BM correctly Flagged as many At-risk students as B-PM did. BM achieved a κ of 0.40.

While we only showed B-PM and BM predictions for the end of the 17 weeks, the models also produced predictions at the end of each week. Flagging students at risk earlier may be more beneficial to a student and an institution's stakeholders since early flagging allows more time for interventions—Section 4.6 reports on the trade-off between timeliness and accuracy.

Table 18. Confusion matrix and summary of BM test set results.

		Safety Score		
		Flagged	**Ignored**	**Total**
Outcome	At-risk	27(27)	19(19)	46
	Safe	22 (12)	102 (112)	124
	Total	49 (39)	121 (131)	**170**
	Outcome	**Precision**	**Recall**	
	At-risk	0.55 (0.69)	0.59 (0.59)	
	Safe	0.84 (0.85)	0.82 (0.90)	
		$\kappa = 0.40$		
		$(\kappa = 0.51)$		

4. Discussion

4.1. Background and Grade

See Figure 2 and Table 12. Bourdieu and Richardson [26] argue that Cultural and Economic Capital regulates the level of success attainable by individuals. The Pearson correlation coefficients for Grade against Quintile and Township School were 0.17 and -0.14, respectively. The Decision Tree used to classify students at risk produced a κ of 0.18 (slight agreement) between the Safety Score and the Outcome. The above relationships between Background and a student's academic output provided evidence for the theories extended by Bourdieu and Richardson [26].

4.2. Extraversion-Level and Grade

See Table 13 and Figure 3. The positive β_0 coefficient of 1.269 signified that the average Grade of students in higher Extraversion levels was higher than the average Grade of students in lower Extraversion levels; while one more post than the last may not result in an additional 1.269 points to a student's Grade record, the average Grade of students who contributed to discussions more frequently, in general, was higher than the Grades of students who posted less often. Although this model accounted for the observed effect on Grade of only one independent variable, Extraversion, the probability (p-value) of Extraversion having no relationship with Grade was 0. This showed a statistical significance of Extraversion as a regressor against student Grade. An increase of 1 in the $E(s)$ correlated with an average Grade increase of 1.269. The Extraversion–Grade relationship was linked to the social science concept of *social capital* for the formation of Academic groups.

4.3. Conscientiousness-Level and Grade

See Table 14 and Figure 4. The β_0 coefficient of $C(s)$ indicated an increase of 1 in Conscientiousness—level was associated with an increase of 5.988 in Grade. Out of the 102 students who ended up at risk of failing their programmes (Grade < 51), 76 had a Conscientiousness level below three. The statistically significant positive correlation between $C(s)$ and $G(s)$ showed that $C(s)$ was a suitable predictor of a student's Outcome.

Hung and Zhang [57] presented a comparable finding; students who accessed course materials 18.5 times or more throughout their programmes obtained a grade of 77.92 out of 100 or higher, while students who accessed course materials more than 44.5 times obtained a grade of 89.62 or higher. Closely related to the above relationship was this study's findings of the correlation between a student's Extraversion level, $E(s)$, and Grade.

4.4. Academic Groups and Social Capital

See Table 16 and Figure 6. The Extraversion–Grade relationship was linked to the social science concept of *social capital*, which was used as a theoretical basis for our formation of Academic groups. Romero et al. [58] obtained a classification accuracy of 60% for the expected grade category of a student (Fail, Pass, Good, Excellent) against their LMS behaviour. Among the features used by Romero et al. [58] was the number of messages sent to a forum (Extraversion level).

Bhandari and Yasunobu [59] and other researchers do not illustrate the quantitative effect of social capital. However, the authors cite that 'an individual who creates and maintains social capital subsequently gains advantage from it [social capital]'. The quantification of the perceived effect of social capital was illustrated by the correlation between the Grade of a student in an Academic group, $Gc_i(h_i)$, and the average Grade of the group, $\mathbb{E}[Gc_i]$. $Gc_i(h_i)$ responded with a statistically significant increase of 0.984 to a $\mathbb{E}[Gc_i]$ increase of 1.

Despite showing different insights and patterns, both the Discussion and Collaboration group methods showed that the *quality* of a student's academic output (Grades) was associated with the quality of the academic output of their social capital. As stated in Section 1.2, a student may choose to *leverage* their social capital (that is available to all students in a cohort) by becoming part of an Academic group.

Our Academic group and Grade relationship, findings such as Romero et al. [58]'s and the above statement by Bhandari and Yasunobu [59] provide evidence for the positive relationship between student success and the accumulation of social capital. This section's work contributed to the theory that:

> *The quality of a student's social capital is the quality of their Academic group's performance.*

Academic Group Size Constraints

Each Discussion and Collaboration group was constrained to a minimum size of three. When the sizes were reduced to two, all linear relationships between the student's Grade and the group's average Grade collapsed and were statistically insignificant. Furthermore, residuals, $Gd_i(s_i) - \hat{G}d_i(s_i)$ (for Discussion–Grade relationships) and $Gc_i(h_i) - \hat{G}c(h)$ (for Collaboration group–Grade relationships) were not normally distributed for the group sizes of two.

4.5. BM and B-PM

Work from cited authors does not discuss how changes in student behaviour relate to changes in student performance. Section 3.6 constructed a temporal e-behaviour machine learning model that yielded a κ of 0.51 against a student's performance.

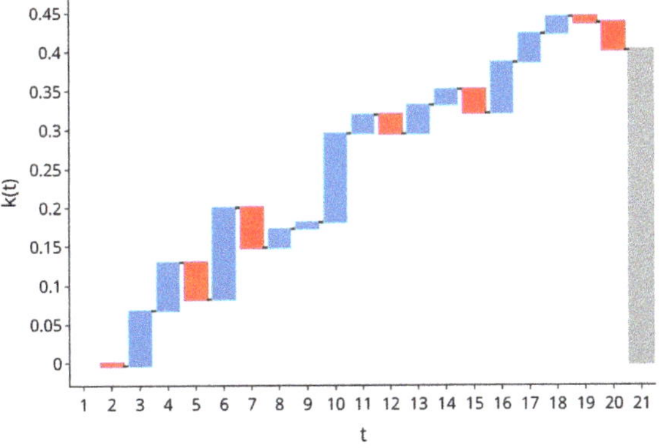

Figure 7. Trade-off between timeliness and accuracy.

4.6. BM and the Trade-Off Waterfall

Figure 7 is a Trade-off Waterfall that illustrates the trade-off between the *benefit of intervention timeliness* and the *cost of intervention success*. Intervention success is determined by the model's accuracy. A red bar represents a decrease in accuracy from one week to the next, and a blue bar indicates an increase. The length of a bar indicates the magnitude of change in accuracy, $\kappa(t)$.

The trade-off between the *benefit of intervention timeliness* and the *cost of intervention success* could help identify whether there were patterns over each year that could inform users about the *optimal* week (t^*) to intervene. If the current year was 2019, then t^* for 2019 (t^*_{2019}) could be determined by either one or a combination of the following factors:

1. t^*_{2019} was chosen to be the t in 2018 that yielded the maximum value of $\kappa(t)$ in 2018.
2. t^*_{2019} was based on *exogenous* considerations determined by the institution's stakeholders. Examples of exogenous considerations were the urgency required for intervention and resources required to make interventions.

For instance, the 2018 Trade-off Waterfall was not available to this study. Therefore the $t^*_{2019} = 17$ used in this cohort's B-PM and BM was based only on the exogenous consideration that interventions should be conducted by week 17.

Practical Benefits and Limitations of the Trade-Off Waterfall

The Trade-off Waterfall was computed after the Outcome of the students was made available. It did not show the week that produced the highest accuracy in real-time, and, in some weeks, the trade-off between accuracy and timeliness was not positive. For example, observe that $\kappa(15) < \kappa(14)$. In exchange for a delayed intervention, BM produced a worse κ score from $\{L(s)_t\}$, which would not have made the delay worthwhile. A similar observation was determined for the delay between weeks 18 and 19. Therefore, there was no way to infer the optimal week to make predictions and interventions in real-time. Instead, the Trade-off Waterfall indicated that:

1. The general trade-off was that $\kappa(t)$ increased as t increased;
2. The trade-off peaked at some point, and, in this case, three weeks before the examination period at $t = 18$. Therefore, it may not be worth waiting for the start of an examination period (such as $t = 21$) before conducting interventions (given the login data of this cohort, different cohorts and different datasets from those presented in this report may produce different peak periods). For example, the Trade-off Waterfall showed that, after $t = 18$, there was no benefit of waiting for an extra one, two or even three weeks to intervene because $\kappa(19), \kappa(20), \kappa(21) < \kappa(18)$.

5. Limitations and Future Work

This research was a study on the methodology that guided the use of machine learning in an academic performance analysis rather than the efficiency and improvement of the algorithms themselves.

Our results might likely differ across contexts, since different data and algorithm configurations can generate several model outputs. The results obtained serve only as proxies for the possible outputs in academic performance research.

In the domain of an LMS system user engagement, there were no formal definitions and standards, analogues or equivalent metrics that proxied a student's e-behaviour and personality from LMS data. We modelled features as well-understood traits to further understand the relationships between behaviour, personality and academic performance.

An unknown in all model outcomes was the presence of causality. For instance, whether e-behaviour had an *effect* on performance was not known. Although the methodology followed aimed to set up conditions for inference, diction such as *tend to correspond with*, and *have relationships with*, instead of *causes*, showed sensitivity to all likelihood of effects from confounding variables.

In encoding performance (student Grade), we used uniform importance across all modules. We did so despite some students' modules accounting for a higher proportion of points towards obtaining a qualification from the University. We did not have access to each student's relative weighting of each module and, therefore, did not account for the differences in module weightings.

5.1. Alternative Formulations of Personalities

An approach to capture various facets of Extraversion and Conscientiousness was attempted. For instance, the *orderliness* facet of Conscientiousness required a metric that modelled the routine or consistency of engagement. Orderliness was modelled by computing the sum of the squared deviations, SS, from each student's mean number of logins. However, the regression model that correlated Grades with SS violated the normality-of-residuals test for normality. Thus, the test for a relationship between SS and Grade was inappropriate under a linear regression model.

Personality tests, as conducted by Costa et al. [60], could be conducted on students in our study. Using personality assessments as an evaluation tool would help understand the extent to which the proxies we developed corresponded to standard personality assessment procedures and could lead to improved proxies. For example, login behaviour did not capture dutifulness as a personality assessment would. Therefore, responses from the assessments could lead to finding proxies that correlated with Conscientiousness in more ways than dutifulness, providing for a fuller assessment of the Conscientiousness proxy since it would be backed by the existing assessment measures.

5.2. Extraversion Levels

Placing the students in Extraversion levels satisfied the OLS assumptions, while regressing each student's post count against their Grade produced statistically significant results; the data's distribution violated OLS assumptions. Hence, the transformation by placing each student into an Extraversion level. Figure 8 shows the Crude Post Count against the Grade of each student. The Residual Normality, Independence, and Homoscedasticity assumptions were violated by the OLS model fitted on the data in Figure 8.

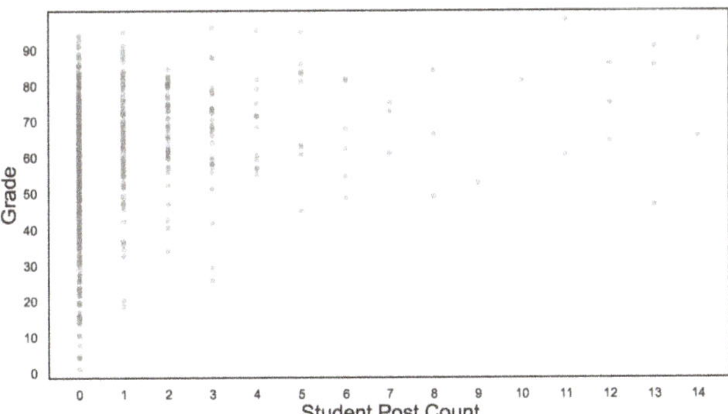

Figure 8. Crude Post Count against Student Grade. There was a positive relationship between Post Count and Grade that was not suitable for a linear OLS fit.

6. Conclusions

We looked at students' Background, behaviour, personality and how these factors were related to student performance. The main difference between our methodology and previous work was that we engineered features from an LMS system. We used these LMS features to act as proxy features for e-behaviours and personality traits as an input to our machine learning models. We then analysed the model outputs and their practical implications. The results demonstrated that a student's background had a lower predictive power of academic performance than their e-behaviour and personality. We found that modelling student behaviours and personality traits required considering how accurately our proposed e-behaviour and personality proxies modelled *true* behaviours and personality traits—we based our models on definitions found in previous literature.

We were able to use Bourdieu's Three Forms of Capital to model social, economic and cultural capital, and the Big Five personality traits to model e-behaviour and personality. From student Background and LMS forum engagement data, Bourdieu's Three Forms of Capital were modelled in the following ways:

- Economic Capital—modelled by the *Financial Assistance*;
- Cultural Capital—modelled by the *Quintile in Province* and *Township School*;
- Social Capital—modelled by Academic groups.

The correlation values for Financial Assistance, Quintile and Township School provided evidence for the authors' argument that Cultural, Economic and Social Capital regulate the level of success attainable by individuals. Cultural and Economic Capital, combined with the Gender feature, performed better than chance at predicting student performance. A student's quality of Social Capital available to them also correlated positively with their academic performance.

We used two of the Big Five personality traits, Conscientiousness and Extraversion, previously found to correlate strongly with performance. Conscientiousness and Extraversion each showed significant predictive performance when used in our MNL models. With these personality features, our e-behaviour classifier achieved better accuracy than without the personality features. The works cited do not discuss how changes in student behaviour relate to changes in their performance. We constructed a temporal e-behaviour model using deep learning that showed an increase in accuracy over time. This e-behaviour learning can be used to flag students at risk at any point throughout their study programmes.

The analyses in this research could be practically useful if they could inform or influence student behaviour. Firstly, a student may find helpful the linear relationship between academic performance and Extraversion. The empirical evidence that a higher Extraversion level is associated with a better academic performance may encourage students to engage in forums more frequently. This evidence may encourage them to engage with the academic content more thoroughly to contribute meaningfully to discussions. Secondly, we showed, using unsupervised cluster learning, that a student's performance was congruent with the performance of their Academic group. The above result is a further motive to action a student into leveraging their social capital by engaging in forums more frequently, since engagement increases their chances of being in an Academic group.

Author Contributions: Conceptualization, S.B.-W.S., R.K. and T.v.Z.; Data curation, S.B.-W.S.; Formal analysis, S.B.-W.S.; Funding acquisition, R.K. and T.v.Z.; Investigation, S.B.-W.S., R.K. and T.v.Z.; Methodology, S.B.-W.S.; Supervision, R.K. and T.v.Z.; Validation, S.B.-W.S.; Visualization, S.B.-W.S.; Writing—original draft, S.B.-W.S., R.K. and T.v.Z.; Writing—review & editing, S.B.-W.S., R.K. and T.v.Z. All authors have read and agreed to the published version of the manuscript.

Funding: The APC was funded by National Research Foundation, South Gate, Meiring Naudé Road, Tshwane, South Africa.

Institutional Review Board Statement: This research filed for a study ethics application that was approved by the Human Research Ethics Committee at the University of the Witwatersrand, Johannesburg, South Africa, 2001. The ethics application included measures imposed on the research methodology to ensure the protection of the student identities and their data's security. This research's clearance certificate protocol number is H19/06/36.

Conflicts of Interest: The funders had no role in the design of the study; in the collection, analyses, or interpretation of data; in the writing of the manuscript, or in the decision to publish the results.

References

1. Richiţeanu-Năstase, E.R.; Stăiculescu, C. University dropout. Causes and solution. *Ment. Health Glob. Chall. J.* **2018**. *1*, 71–75.
2. Wright, D.; Taylor, A. *Introducing Psychology: An Experimental Approach*; Penguin Books: London, UK, 1970.

MDPI

St. Alban-Anlage 66

4052 Basel

Switzerland

Tel. +41 61 683 77 34

Fax +41 61 302 89 18

www.mdpi.com

Applied Sciences Editorial Office

E-mail: applsci@mdpi.com

www.mdpi.com/journal/applsci